Phytochemistry of *Piper betle* Landraces

Phytochemical Investigations of Medicinal Plants

Series Editor:
Brijesh Kumar

Phytochemistry of Plants of Genus *Phyllanthus*
Brijesh Kumar, Sunil Kumar and K. P. Madhusudanan

Phytochemistry of Plants of Genus *Ocimum*
Brijesh Kumar, Vikas Bajpai, Surabhi Tiwari and Renu Pandey

Phytochemistry of Plants of Genus *Piper*
Brijesh Kumar, Surabhi Tiwari, Vikas Bajpai and Bikarma Singh

Phytochemistry of *Tinospora cordifolia*
Brijesh Kumar, Vikas Bajpai and Nikhil Kumar

Phytochemistry of Plants of Genus *Rauvolfia*
Brijesh Kumar, Sunil Kumar, Vikas Bajpai and K. P. Madhusudanan

Phytochemistry of *Piper betle* Landraces
Vikas Bajpai, Nikhil Kumar and Brijesh Kumar

For more information about this series, please visit: https://www.crcpress.com/Phytochemical-Investigations-of-Medicinal-Plants/book-series/PHYTO

Phytochemistry of
Piper betle Landraces

Vikas Bajpai, Nikhil Kumar and
Brijesh Kumar

CRC Press
Taylor & Francis Group
Boca Raton London New York

CRC Press is an imprint of the
Taylor & Francis Group, an **informa** business

First edition published 2020
by CRC Press
6000 Broken Sound Parkway NW, Suite 300, Boca Raton, FL 33487-2742

and by CRC Press
2 Park Square, Milton Park, Abingdon, Oxon, OX14 4RN

ISBN: 978-0-367-85965-7 (hbk)
ISBN: 978-0-367-49971-6 (pbk)
ISBN: 978-1-003-01605-2 (ebk)

Typeset in Times
by codeMantra

Contents

List of Figures ix
List of Tables xi
Preface xiii
Acknowledgments xv
Authors xvii
List of Abbreviations and Units xix

1 Introduction **1**
  1.1 *Piper betle*: The Plant and Its Distribution 4
  1.2 Botanical Description of *Piper betle* 5
  1.3 Economic Potential of Leaves 6
  1.4 Nutritive Value of Leaves 6
  1.5 Morphoanatomical Studies of *Piper betle* L. Landraces 6
  1.6 Etymology 7
  1.7 Cultivation of *Piper betle* 7
  1.8 Brief Review on the Chemistry of *Piper betle* Leaves 9
  1.9 Brief Review on Pharmacology and Biological Activity
     of *Piper betle* 10
  1.10 Traditional Uses of *Piper betle* 11
    1.10.1 Antibacterial Activity 12
    1.10.2 Antidiabetic Activity 12
    1.10.3 Antifertility Activity 13
    1.10.4 Antiinflammatory and Antiallergic Response 13
    1.10.5 Antimalarial Activity 13
    1.10.6 Antioxidant Activity 14
    1.10.7 Insecticidal Activity 14
    1.10.8 Immunomodulatory Activity 14
    1.10.9 Chemopreventive and Anticancer Activity 14
    1.10.10 Cholinomimetic Effect 15
    1.10.11 Neuropharmacological Profile 15
    1.10.12 Platelet Inhibition Activity 15
    1.10.13 Protective and Healing Activity 15
    1.10.14 Radioprotective Activity 16
    1.10.15 Antidermatophytic Activity 16

1.10.16 Antihypercholesterolemic Activity   16
1.10.17 Antinociceptive Activity   16
1.10.18 Gastroprotective Activity   17
1.10.19 Antiasthmatic Effect   17
1.10.20 Effect on Thyroid Function Activity   17

2 Metabolite Profiling of *Piper betle* Landraces by Direct Analysis in Real Time Mass Spectrometric Technique   19
2.1   Plant Material and Chemicals   19
2.2   Extraction and Sample Preparation   20
2.3   Preparation of Samples for Direct Analysis in Real Time Mass Spectrometric Analysis   23
2.4   Analysis Conditions   23
2.5   Multivariate Statistical Analysis   24
2.6   Method Validation   24
2.7   Optimization of Direct analysis in Real Time Mass Spectrometric Conditions   25
2.8   Discrimination of *Piper betle* Landraces   29
2.9   Differences in *Piper betle* Male and Female Leaf Metabolite Profile   31
2.10 Phytochemical Profiling to Classify the Therapeutic Potential of *Piper betle* Landraces   35
2.11 Discrimination of the *Piper betle* Landraces from Different Geographical Origin   38

3 LC-MS Analysis of *Piper betle* Leaf and Evaluation of In Vitro Antimicrobial Activity   41
3.1   Plant Material, Reagents and Chemicals   41
3.2   Extraction and Sample Preparation   42
3.3   Preparation of Standard Solution   42
3.4   Instrumentation and Analytical Conditions   43
3.5   Optimization of Ultrahigh Performance Liquid Chromatography–Tandem Mass Spectrometric Conditions 44
3.6   Qualitative Analysis   44
      3.6.1 Identification of Propenyl Phenols and Their Derivatives   48
      3.6.2 Identification and Characterization of Other Phytoconstituents   49
3.7   Validation Parameters of Quantitative Analysis   50
      3.7.1 Linearity, Limit of Detection and Limit of Quantification   50
      3.7.2 Precision, Stability, and Recovery   51

3.8   Quantitative Analysis   51
3.9   Principal Component Analysis   53
3.10 Antimicrobial Activity   54
3.11 Antimicrobial Activity Evaluation   54

4 Conclusions   **57**

References   59
Index   67

List of Figures

1.1	Distribution of *Piper betle* in different regions in the world	5
1.2	Cultivation of *Piper betle*	8
2.1	DART mass spectrum of Saufia	29
2.2	DART mass spectrum of Deshi	30
2.3	Validated PCA score plot of *Piper betle* landraces	32
2.4	Identification and validation of gender in *Piper betle* using PCA	34
2.5	Identification of gender of thirty eight *Piper betle* landraces	35
2.6	Tree view of therapeutic potential cluster for twenty one *Piper betle* landraces	38

List of Tables

1.1 Some important cultural plants of India with their common names and botanical equivalents   2

1.2 Common names of *Piper betle* in different region   3

2.1 *Piper betle* landraces with their collection or procurement sites   21

2.2 Optimization of DART-MS parameters for the analysis of *Piper betle* leaf in positive and negative ionization modes   26

2.3 DART-MS based Identification of Phytochemicals in *Piper betle* leaf   28

2.4 DART mass spectral data of *Piper betle* landraces, Bangla (1), Desawari (2), Deshi (3), J. Green (4), J. White (5), Kalkatiya (6), Mahoba (7), and Saufia (8)   31

2.5 Gender determination of unknown *Piper betle* on the basis of peak abundance   34

3.1 The [M+H]+ ions, MS/MS fragment ions and UV absorption maxima for compounds identified from leaf extracts of *Piper betle* by UHPLC-ESI-MS/MS experiment   45

3.2 The content (mg/g) of three analytes in thirteen *Piper betle* landraces   52

3.3 Minimum inhibitory concentration (MIC) in μg/mL of ethanolic leaf extracts of *Piper betle* landraces against bacteria and fungi   55

Preface

Bioactive secondary metabolites of medicinal plants are considered to be a fundamental source of medicine for the treatment of a range of diseases in the modern medical system. Historically, prior to the advent of modern medical system, humans depended on natural products, especially the plants, to promote and maintain good immune system to fight illness, discomfort and diseases. Plants are the foundation on which the whole living world thrives and survives. Without the extraordinary variety of living organisms including the world of plants, animal life would not have survived and our planet would have been a barren and lifeless desert. India is one of the world's important biodiversity centers with the existence of more than 45,000 diverse plant species and out of these, around 15,000–20,000 species are medicinally important and nearly 7000–7500 species are being used by folklore and traditional practitioners. Thus, India has a rich heritage of medicinal plants constituting the basis of indigenous systems of medicine such as Ayurveda, Siddha, Unani, Homoeopathy and Naturopathy.

Piper betle, primarily used as a leaf, is one of the most important plants of culture and considered to be a post Vedic plant, which has become an integral part of the rituals transcending cultural divides. *P. betle* is a Pan Asiatic plant, and as per estimate, approximately 600 million people use it every day in one form or the other. The Indian system of medicine Ayurveda recognized its importance, and thirteen properties are listed in Ayurvedic texts. Considering the magnitude of its use, it was also known as the green gold in India and other southeast Asian countries. Based on the work done in India over the past 50 years, around 150 local landraces (naturally occurring stable variants within a species) were recognized. Similar diversity is also expected in other regions where it is grown or available in the wild. Differences in the bioactivity among the various *P. betle* landraces are supposed to be due to the difference in their chemical constituents. Reliable phytochemical fingerprint or profile of these plants validates the diversity.

P. betle occupies a unique position due to its multiple uses including medicinal uses and also the quantity of its consumption. In view of this, it may be of great significance to know not only the phytochemical composition of the plant but also its pharmacological effects. The phytochemical analysis provides the relationship between the composition of complex and variable mixtures of plant derived medicines and their biological effects. One of the trends in modern analytical research is the development of new strategies, which must be simple but capable and robust to reliably screen all the major compounds in the sample without any tedious sample preparation process. Direct analysis in real time mass spectrometer (DART-MS) is one such technique that offers sample analysis without any sample preparation, and it can analyze phytochemicals in intact plant materials. Further, mass spectrometry coupled with liquid chromatography is very useful in the identification of medicinally active ingredients in complex mixtures of plant extracts, as it involves prior separation before mass spectrometric identification. Mass spectrometers having highly sensitive mass analyzers such as time of flight (TOF) deliver high resolution mass spectrometric data, which are useful in the identification of chemical compounds differing in their exact masses. Mass spectrometry offers great selectivity and sensitivity, and with the separation power of liquid chromatography, it allows simultaneous structural analysis and quantitation of phytochemicals present at the sub-ppm level in complex matrices of plant extracts.

This book deals with the development of a new approach for metabolite profiling to understand the extent of variability in the metabolites due to landraces, gender and geographical location without the classical sample preparation by the combination of direct analysis in real time-time of flight-mass spectrometric method and chemometric analysis. A total of 63 *P. betle* landraces were analyzed using DART-MS. In *P. betle* leaf, more than fifteen compounds were identified and characterized based on exact mass measurements and MS/MS patterns. Different landraces showed characteristic differences in the secondary metabolites. The marker compounds were identified for differentiating landraces and geographical variations in the leaf. Based on peak abundance and principal component analysis it was also possible to identify plant gender in this dioecious plant which does not flower at many places where it is cultivated.

31/12/2019   **The Authors**

Acknowledgments

We owe our gratitude to a number of people who inspired us in their own ways to the path of knowledge, work ethics and the required determination to conceive of this book and complete it.

We would like to thank all the colleagues who revised several sections of the book and provided many useful suggestions. This book would not have been possible without the support and help of Mr. M.P.S. Negi. The characterization and quantitation in this book is based on the input from Dr. Renu Pandey of our research group. We are also thankful to Dr. K. P. Madhusudanan, whose encouragement, guidance and support from the beginning to the final stage enabled us to develop understanding of the subject. His constructive criticism and warm encouragement made it possible for us to bring the work to its present shape. We are indebted to him for shaping our thoughts. We express a deep sense of gratitude to the Director, CSIR-Central Drug Research Institute (CDRI), Lucknow, India, for his support.

Authors

Dr. Vikas Bajpai completed his PhD from the Academy of Scientific and Innovative Research (AcSIR), New Delhi, India, and carried his research work under supervision of Dr. Brijesh Kumar at CSIR-Central Drug Research Institute, Lucknow, India. His research includes development and validation of LC-MS/MS methods for qualitative and quantitative analysis of small molecules of Indian medicinal plants.

Dr. Nikhil Kumar is a plant physiologist by training and has worked in different capacities for about 30 years and finally superannuated from CSIR-National Botanical Research Institute (NBRI), Lucknow, India. He has worked on one of the very important cultural plants of India (including southeast Asia) *Piper betle*, which effectively broadened his understanding as how plants contributed in the development of human skills. He has published more than fifty research papers in national and international journals. He brought to focus the aspects of dioecy in *Piper betle* and *Tinospora cordifolia* and its possible functional implications in adaptation and biological activities.

Dr. Brijesh Kumar is a Professor (AcSIR) and Chief Scientist of Sophisticated Analytical Instrument Facility (SAIF) Division, CSIR-Central Drug Research Institute (CDRI), Lucknow, India. Currently, he is facility in charge at SAIF Division. He has completed his PhD from CSIR-CDRI, Lucknow (Dr. R. M. L Avadh University, Faizabad, UP, India). He has to his credit 7 book chapters, 1 book, and 145 papers in reputed international journals. His current area of research includes applications of mass spectrometry (DART MS/QTOF LC-MS/4000 QTrap LC-MS/ Orbitrap MSn) for qualitative and quantitative analyses of molecules for quality control and authentication/standardization of Indian medicinal plants/parts and their herbal formulations. He is also involved in identification of marker compounds using statistical software to check adulteration/substitution.

List of Abbreviations and Units

°C	degree Celsius
μg	microgram
μL	microliter
AP	Andhra Pradesh
APC	allylpyrocatechol
As	Assam
bdl	below detection level
CA	cluster analysis
CAD	collision activated dissociation
CHV	chavicol
CID	collision induced dissociation
DART	direct analysis in real time
DP	declustering potential
EP	entrance potential
ESI	electrospray ionization
FA	factor analysis
GS	gas
HPLC	high performance liquid chromatography
HRMS	high resolution mass spectrometry
ICH	International Conference on Harmonization
KL	Kerala
KN	Karnataka
LC-MS	liquid chromatography–mass spectrometry
LOD	limit of detection
LOQ	limit of quantitation
mg	milligram
MH	Maharashtra
MIC	minimum inhibitory concentration
ml	mili liter
MMU	milli mass unit
MRM	multiple reactions monitoring

MS	mass spectrometry
PB	*Piper betle*
PAD	photodiode array detector
PC	principal component
PCA	principal component analysis
PEG	polyethylene glycol
PPM	parts per million
RAPD	randomly amplified polymorphic DNA
SD	standard deviation
TOF	time of flight
UHPLC	ultrahigh performance liquid chromatography
UP	Uttar Pradesh
UV	ultraviolet
WB	West Bengal
WHO	World Health Organization

Introduction

One of the human traits, the ability to use natural resources such as plants and animals, contributed to the evolution from hunter gatherer to modern humans. Human history is the history of the ability to use and transform the natural resources for their own benefit. Everything around us today is the manifestation of this unique human ability. Green plants, as we know today, are the first life forms, which in many ways transformed the earth and prepared the platform for the evolution and survival of other life forms and in that series, humans were the last to arrive. The green plants from single celled microscopic algae to big vascular plants such as redwoods or giant sequoia, the tallest living plant reaching heights of more than 350 feet (107 m), show the extent of variability in the plant world. Plant taxonomists working in different parts of the world have cataloged all the terrestrial vascular plants and the number is around 3, 80,610 species. Nearly 10% of these species are with known direct or indirect human uses, and about 17,000 species are known for their different medicinal uses (Nayar, 1987), the number is somewhat less based on different known traditional medicinal systems. In India, many sections of the population widely use medicinal plants, and it has been assessed that, in all, more than 7,500 plant species are used by many ethnic communities (AICEP, 1994; Anthropological Survey of India, 1994;). In China, more than 7,000 plant species have been recognized as medicinal plants (Medicinal Plants in China: A Selection of 150 Commonly Used Species. WHO Regional Publications: Western Pacific Series No. 2, 1989). Besides this, there is yet another group of plants, known as cultural plants, which is a collection of useful plants from different subgroups such as food, fiber, spices, and medicine. The important cultural plants of India are listed in Table 1.1 and the list is indicative as the numbers may vary.

Family Piperaceae contains approximately 3,600 species distributed in 13 genera (Arias et al., 2006). The genus *Piper* belongs to family Piperaceae with approximately 2,000 known species with worldwide pan tropical distribution (Quijano Abril et al., 2006) and is one of the 20 most species rich genera of flowering plants (Frodin, 2004). *Piper* is an important genus in the Indian medicine system or the traditional system of medicine due to its well known therapeutic activities. Among all the documented and well recognized species

1

TABLE 1.1 Some important cultural plants of India with their common names and botanical equivalents

COMMON NAME (BOTANICAL NAME)	COMMON NAME (BOTANICAL NAME)
Doorva (*Cynodon dactylon*)	Gudhal (*Hibiscus rosa sinensis*)
Barley (*Hordeum vulgare*)	Madar (*Calotropis gigantea*)
Coconut (*Cocos nucifera*)	Aam (*Mangifera indica*)
Til (*Sesamum indicum*)	Datura (*Datura stramonium*)
Neem (*Azadirachta indica*)	Shami (*Prosopis cineraria*)
Bhang (*Cannabis indica*)	Guduchi (*Tinospora cordifolia*)
Cotton (*Gossypium* sp.)	Supari (*Areca catechu*)
Banana (*Musa* sp.)	Kusha (*Desmostachya bipinnata*)
Tulsi (*Ocimum sanctum*)	Paan (*Piper betle*)
Kaner (*Cascabela thevetia*)	Bilva (*Aegle marmelos*)
Chandan/Sandal (*Santalum album*)	Haldi (*Curcuma longa*)
Parijat (*Nyctanthes arbortristis*)	Banyan (*Ficus benghalensis*)
Kamal (*Nelumbo nucifera*)	Rudraksha (*Elaeocarpus ganitrus*)
Marigold (*Tagetes* spp.)	Peepal (*Ficus religiosa*)

of the genus *Piper* are *Piper nigrum, Piper betle, Piper longum, Piper chaba* etc. The presence of several unique phytochemicals as secondary metabolites is accountable for the traditional and commercial uses of *Piper* species. Different class of compounds such as lignans, amides, long chain esters, terpenoids, amides, benzoic acids, chromenes, terpenes, phenylpropanoids, phenolics and a series of alkaloids, steroids, pyrones, chalcones and flavonoids are reported to be found in *Piper* species (Parmar et al., 1998; Valentão et al., 2010). Alkaloids and phenolics are the most important class of compounds found in plants of various *Piper* species. Piperamided such as piperine (1-piperoylpiperidine) and piperbetols such as chavibetol from *Piper* species has shown many pharmacological activities (Khajuria et al., 1998; Dorman and Stanley, 2000; Sunila and Kuttan, 2004; Lee et al., 2005; Desai et al., 2008; Mujumdar et al., 1990; Dhuley et al., 1993; Daware et al., 2000; Bajad et al., 2001a; Bajad et al., 2001b). The genus *Piper* has mostly commercial and economic significance for pharmaceutical and beverage industries. In terms of phylogenetic position, Piperaceae is among the diverse assemblage of dicots termed "paleoherbs," and its vegetation resembles with monocots in certain vegetative functions inclusive of adaxial prophyll and scattered vascular bundles. *Piper* is one of the most diversified genera in the family Piperaceae and occupies a basal lineage position within the angiosperm group. *Piper* includes timber and herbs that can be discovered in humid or wet places all over the world, mainly in tropical regions of both the hemispheres (Parmar et al., 1997; Chopra and Vishwakarma, 2018).

Paan/betelvine (*P. betle* Linn.) is a tropical species that is also closely related to the common pepper, and it belongs to the family Piperaceae. The wide majority of pepper plants are covered within the two foremost genera: one is *Piper* that contains approximately 2,000 species and other is *Peperomia* with approximately 1,600 species. The population of the plants in the Piperaceae family is varied from small trees, shrubs to herbs. The described distribution of this group is mainly as pantropical. The best known species from *Piper* genera is *P. nigrum*, known for production of most of the peppercorns, which is basically used as spice, together with black pepper, although its relatives in the family include many other spices. The family name Piperaceae is likely to be derived from the term *pippali*, a Sanskrit word that was used to describe long peppers. *P. betle* (PB) is a pan Asiatic plant grown in the tropics and part of the rich culture of this region. The oldest record of use of PB came from the fossil remains of cavemen in Thailand dating circa 5000 BCE (Pradhan et al., 2013). The cultural impact of this plant can be understood by the fact that it is known by different local names (Table 1.2). PB leaves are known to Indian culture from centuries and are specially included in all rituals right from birth to death. According to estimates, nearly 600 million people consume it daily in customized forms. Based on the magnitude of consumption, it only ranks below coffee and tea. PB leaves and inflorescence are introduced into

TABLE 1.2 Common names of *Piper betle* in different region

S. NO.	LANGUAGE	COMMON NAME
1	Sanskrit	Nagavallari, Nagini, Nagavallika, Tambool, Saptashira, Mukhbhushan, Varnalata
2	Malay	Sirih, Sirih melayu, Sirih cina, Sirih hudang, Sirih carang, Sirih kerakap
3	English	Betle, Betle pepper, Betelvine
4	Tamil	Vetrilai
5	Telugu	Nagballi, Tamalapaku
6	Hindi	Paan
7	Gujarati	Nagarbael
8	Marathi	Nagbael
9	Bengali	Paan
10	Semang	Serasa, Cabe
11	Jakun	Kerekap, Kenayek
12	Thai	Pelu
13	Sakai	Jerak
14	Javanese	Sirih, Suruh, Bodeh

the betel quid (BQ) for chewing. The BQ chewing practice has social sanction and credited with many medicinal effects including digestive, simulative, carminative and aphrodisiac (Prabhu et al., 1995; Row and Ho, 2009) however, its prolonged abuse leads to several problems. Phytochemical known as allylpyrocatechol from PB leaves exhibited promising effect against obligate oral anerobes liable for halitosis (Ramji et al., 2002; Row and Ho, 2009; Nath et al., 2011).

1.1 *PIPER BETLE*: THE PLANT AND ITS DISTRIBUTION

PB is a shade loving climber, widely cultivated in India in the tropical and subtropical regions with mild and short winter spell. Since the plant occupied a very important place in the Indian society, its supply to different regions was required round the year. Given the climatic conditions, it was not possible to cultivate it in all the regions of India despite pan India use. This led to the development of cultivation practices, which made it possible to cultivate PB in those areas where its natural growth was not supported by the prevailing climatic conditions. Thus, a whole set of procedures for controlled cultivation was evolved and successful cultivation of PB under unfavorable conditions became a reality (Kumar, 1999). PB cultivation in other southeast Asian countries with a mild weather throughout the year does not require any special arrangements unlike India where the plant has a small window for growth especially in the subtropics and often suffers crop damage due to low temperature/frost during winters (Tripathi et al., 2000). Though all the plant parts are used, the leaf is the most important and used fresh or after a period of storage, which leads to loss of chlorophyll and the pungency in comparison with the fresh leaves. Due to xeromorphic character of the leaf, it is amenable to storage and also able to retain the aroma (Seetha Lakshmi and Naidu, 2010). Since the crop is essentially grown for the leaves, it is also called green gold.

PB is essentially a vegetatively propagated plant through stem cuttings; propagation through seeds is not practiced by the growers, as the flowering and seed formation is confined to few regions where the annual growth cycle is not interrupted by climatic conditions (hot and cold). The other practical reason is the economics as the seed raised plants will take several months to establish and lead to net loss in the annual yield cycle. The other advantage of vegetative propagation is the stability in characters such as leaf yield, texture, aroma, and other organoleptic properties. As an outcome of surveys conducted over several years in India, the occurrence of approximately 150 PB landraces has been documented. The difference between a variety and landrace lies in the fact that whereas a variety is an outcome of the anthropogenic intervention in sexual breeding followed by selection, a landrace is an outcome of stable spontaneous mutation confined to a particular geographical location (Balasubrahmanyam et al., 1994; Samantaray et al., 2012). Obligate vegetative propagation and domestication over hundreds of years and spontaneous stable mutations have led to the emergence of more than 150 landraces in PB. Many of these PB landraces differ from each other in morphological, postharvest shelf life, organoleptic properties and DNA polymorphism (Samantaray et al., 2012; Verma et al., 2004; Ranade et al., 2011). Landrace names carry a strong local stamp in their etymology, and many a time, a certain landrace may be named differently in different regions due to lack characterization; however, in recent years, some efforts have been made in this direction (Priya, 2011; Ranade et al., 2011).

1.2 BOTANICAL DESCRIPTION OF *PIPER BETLE*

P. betle (Sanskrit—Saptashira) is a perennial, evergreen, dioecious, semiwoody climber, shade loving climber widely grown in the tropical humid climate. The plant has alternate, glossy deep green heart shaped smooth, shinning, and long stalked leaves (Figure 1.1 with pointed apex. It has five to seven ribs arising from the base, minute flowers, and one seeded spherical small berries. PB is a dioecious plant bearing male and female flower on different plants. The edible leaves of this plant possess an aromatic flavor and strong pungent smell and are medicinally and economically important. Typical analysis of fresh leaf showed moisture content (85.4%), protein (3.1%), fat (0.8%), minerals (2.3%),

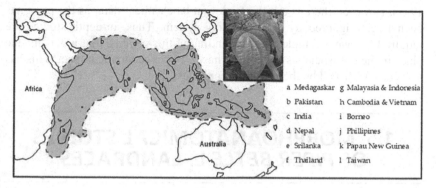

FIGURE 1.1   Distribution of *Piper betle* in different regions in the world. (Reproduced from Kumar et al., 2010b with permission from Indian Academy of Sciences.)

fiber (2.3%), and carbohydrates (6.1%). Its minerals and vitamin contents include calcium, magnesium, iron, carotene, niacin, riboflavin, thiamine, and vitamin C. Traditionally, it is believed that the leaves with lateral veins making a broad loop rejoining the midrib are medicinally more potent than those leaves where this is not prominent (Arawwawala et al., 2014; Shah et al., 2016).

1.3 ECONOMIC POTENTIAL OF LEAVES

The vast economic potential of the PB crop can be adequately established by the fact that globally around 500–600 million people consume it on a daily basis (Kumar et al., 2010b; Sripradha, 2014), and a maximum number of users are from Indian subcontinent (Guha, 2006; Jane, 2014). As per estimates, in India, this crop provides a national income to the tune of INR 6,000–7,000 million annually. Due to globalization, better trade facilities and population migration, the trade in PB leaves have increased, and the leaves are exported to several gulf countries, Europe and parts of the new world (Guha, 2006).

1.4 NUTRITIVE VALUE OF LEAVES

According to folklore, six PB leaves with slaked lime is said to be equivalent to about 300 mL of cow's milk and also provide vitamins and useful minerals. The leaves also contain several digestive enzymes and significant amount of all the essential amino acids (CSIR, 1969; Gopalan, 1984; Guha and Jain, 1997). However, there is a lack of relevant biochemical data, as the plant has been largely ignored by the prevailing system. Thus, in-depth studies are required to have a complete understanding of this plant so that it gets its due place in the user societies and also to have a better economic impact. This has to be accomplished by the core user countries.

1.5 MORPHOANATOMICAL STUDIES OF *PIPER BETLE* L. LANDRACES

Seetha Lakshmi and Naidu (2010) studied the comparative morphoanatomical structures of ten common landraces of *P. betle* available in India.

The ten landraces showed some structural similarities. Four layered upper and two layered lower epidermis was observed in all the landraces of PB studied. Crystals and oil reserves were found in the epidermal cells. The Kapoori Tuni landrace has more stomatal and trichome frequency. Multicellular tector trichomes were seen on the abaxial face of midrib. Presence of parenchymatous bundle sheath was seen in all the landraces. The developments of tracheoids idioblasts from the vessel elements have been noticed. These characteristics are typical xeromorphic anatomy of leaves, which could preserve the aroma and shelf life, longevity of betle leaves. In this study, a clear distinction was observed between Kapoori Tuni and other landraces.

1.6 ETYMOLOGY

The genus *Piper* has been named after one of the important plants of the genus called "pippali" also known as pepper. The specific epithet for species betle is from the Malayalam vernacular name, vettila, for the masticatory leaves. It helps in food digestion and in normalizing the digestive tract.

1.7 CULTIVATION OF *PIPER BETLE*

Cultivation of PB requires a porous, fertile soil on upland site, as the plant is very sensitive to water logging; a typical highland cultivation. In India, there are three types of cultivation systems with varying degree of anthropogenic interventions: natural, partially controlled, and fully controlled cultivation (Figure 1.2). The controls are essentially for the regulation of light, temperature and humidity and in that sense represent the first anthropogenic intervention in terms of raising a crop under unfavorable conditions (Kumar, 1999; Jane, 2014). In Bengal, farmers prepare a garden called a *barouj* (vernacular Bangla) fully controlled facility for cultivation, similar to what is done in subtropical India where it is also known by various names like *bareja/bheet*. The soil is prepared well and laid out into furrows of 10–15 m length, 75 cm width, and 75 cm depth. Oil cakes and well rotten farmyard manures are mixed well with the topsoil of the furrows. Single node vine cuttings are planted after proper dressing from February to June, depending on the location before the onset of monsoon season. The planting is done in parallel rows about two feet apart to avoid mutual shading and the saplings are twined around upright sticks of split bamboo and reeds as the plants start growing.

FIGURE 1.2 Cultivation of *Piper betle.* (Reproduced from Kumar, 1999 with permission from National Institute of Sciences of India.)

The barejas are ingeniously designed special growth enclosures, use locally available plant materials for erection and sustenance of the structure. The barouj/bareja/bheet is a very big hut like structure supported by bamboo sticks and wooden poles as frame with thatching comprising any suitable material from paddy straw, *Saccharum munja* to coconut leaves. Some of the most significant features of *bareja* (Figure 1.2 A-H) are as follows:

- Light regulation within the enclosures is the most important activity.
- Sun flecks show the extent of light modulation.
- Partial regulation of temperature and humidity.
- Season mild winters.
- Fully grown vines within bareja.
- Mean winter temperature–maximum: 20°C–25°C; minimum: 5°C–10°C (Site Mahoba, UP, India (location in degrees: coordinates))
- Despite top cover at times, low temperature/frost causes crop damage in the subtropical regions

The *barouj/bareja/bheet* is a very big hut like structure supported by bamboo sticks and wooden poles as frame with thatching comprising any suitable material from paddy straw to coconut leaves. Since this technology was developed by the people, it has inherent flexibility/diversity as per the climatic requirement and the availability of natural resources for raising the structure.

PB being a native of tropics has inherent sensitivity to cold. It was observed that some of the landraces are more sensitive to cold than others (Tripathi et al., 2000). The visible effect of cold was on the chlorophyll level in the leaf. It was observed that the first enzyme in chlorophyll degradation, chlorophyllase was more active in those leaves which showed greater chlorophyll loss like Madras Paan Kapoori and Kapoori Vellaikodi and less active in Deshi Bangla, Kaker and Bangla Mahoba showing lesser chlorophyll loss (Gupta et al., 2012). This response was more in male landraces than in female landraces. Molecular studies using different DNA markers have shown that the PB landraces can be segregated on the basis of gender (Ranade et al., 2002 and Samantaray et al., 2012) into two broad groups; male and female landraces. Thus, the phenomenon of dioecy must be considered in all the PB studies.

1.8 BRIEF REVIEW ON THE CHEMISTRY OF *PIPER BETLE* LEAVES

The essential oil of PB contains mostly terpenes and phenols. The characteristic flavor of betel leaf is due to the presence of betel phenols. The active constituents of PB leaves are primarily a class of allylbenzene compounds and betel phenols (Rimando et al., 1986). The 1,8-cineole, cadinene, camphene, caryophylline, limonene, pinene, etc. are the terpenoids, and allyl-pyrocatechol, carvacrol, safrole and chavibetol are the major phenols found in PB leaves. Their acetates are also commonly found such as chavibetol acetate, allylpyrocatechol monoacetate, eugenol acetate and allylpyrocatechol diacetate. Chemically, the PB leaves also contain arecoline, piperitol, piperbetol, isoeugenol, safrole, anethole, β-sitosteryl palmitate, dotriacontanoic acid, β-sitosterol, tritriacontane, stearic acid, hydroxychavicol, cepharadione, piperine, piperlonguminine, allyl diacetoxy benzene and methyl piperbetol, p-cymene, α-terpine, α-terpinol, terpinyl acetate, stearaldehyde, ursonic acid, 3β-acetyl ursolic acid, piperol A and piperol B. However, specific emphasis has been placed on chavibetol (betel phenol: 3-hydroxy-4-methoxyallylbenzene). PB phenols also comprise chavicol (p-allyl-phenol: 4-allyl-phenol), estragole (p-allyl-anisole: 4-methoxy-allylbenzene), eugenol (allylguaiacol: 4-hydroxy-3-methoxy-allylbenzene: 2-methoxy-4-allyl-phenol), methyl eugenol (eugenol methyl ether: 3,4-dimethoxy-allylbenzene), and hydroxycatechol (2,4-dihydroxy-allylbenzene). The leaf is also rich in carotenes, ascorbic acid, tannins and sugars and also contains mutagenic compounds, *N*-nitrosopiperidine, *N*-nitrosopyrrolidine, and *N*-nitrosomorpholine.

The study of the volatile content of PB is important because of the bioactivity and organoleptic features that these phytocompounds exhibit (Rawat et al., 1987, 1989). Volatile metabolites show significant roles in plants, as they act as repellents for herbivores and pathogens and help in determining plant communications. They can also be determinant in fascinating pollinators and have the capability to defend against oxidative damage.

1.9 BRIEF REVIEW ON PHARMACOLOGY AND BIOLOGICAL ACTIVITY OF *PIPER BETLE*

ताम्बूलवल्ली तम्बूली भागिनी नागवल्लरी । ताम्बूलं विशदं रूच्यं तीक्ष्णोष्णं तुवरं सरमू ॥११॥
वश्यं तिक्तं कटु क्षारं रक्तपित्तकरं लघु । बल्यं श्लेष्मास्यदौर्गन्ध्यमलवातश्रमापहमू ॥१२॥

(भा. प्र. गुडूच्यादिवर्ग ११-१२)

अथ भवति नागवल्ली तम्बूली फण लता च सप्तशिरा पर्णलता फनिवल्ली भुजगलता भक्ष्यपत्री च

(राज निघंटू)

ताम्बूलं कटुतिक्तमुष्णमधुरं क्षारं काषायान्वितं वातघ्नं कफनाशनं कृमिहरं दुर्गन्धिनिर्णाशनम्

वक्त्रस्याभरणं विशुद्धिकरणं कामाग्निसंदीपनं ताम्बूलस्य सखे! त्रयोदश गुणाः स्वर्गेऽपि ते दुर्लभाः

(ताम्बूल मन्जरी)

The first two quotes are from standard Ayurvedic texts and the third one from a text written in the early nineteenth century. *P. betle* leaf has acrid, astringent, anthelmintic, aphrodisiac, aromatic, bitter, carminative, desiccative, exhilarant, expectorant, febrifuge, stomachic, sweet and laxative properties. It is valuable in throat and chest ailments and halitosis and is also the best mouth freshener. The word Paan for betel leaf has its origin claimed to be derived from Prana life, and it was used as offering to deities and part of all the religions (Hinduism, Buddhism, and Jainism except Sikhism) evolved in India. It was determined that the mixture of betel leaf, clove, and cardamom is beneficial in oral hygiene and controls the oral microbial population due to the synergistic antimicrobial action (Shitut et al., 1999; Bissa et al., 2007; Patra et al., 2016). Dental caries had a chronic endogenous contamination due to the normal oral commensal flora. The bacteria primarily liable for dental decay in humans are *Streptococcus mutans*. Streptococci belong to four species group:

mutan, salivarius, anginosus and mitis. In addition to *S. mutans, Lactobacillus acidophilus* bacteria likely play a minor role in acid production in the plaque (Bissa et al., 2007; Patra et al., 2016). It was shown that betel chewing has the effect on imperative and autonomic fearful systems (Chu, 2001; Patra et al., 2016). The anticariogenic effect of crude PB extract was assessed by its effect on salivary pH and it was found to resist pH change (Varunkuma et al., 2014; Patra et al., 2016).

1.10 TRADITIONAL USES OF *PIPER BETLE*

The leaf extract of the plant has been traditionally used in curing inflammation and infections of the respiratory tract, cough, diphtheria, dyspnea, hysteria, indigestion, as well as general and sexual debility. The leaf juice is given systemically to treat cough and indigestion in infant and children. It also shows antimalarial, antifungal and antibacterial effects, insecticidal activities, antioxidant activity, antidiabetic effect, gastroprotective effect, antinociceptive effect, cytotoxic effect and antiplatelet activity. Stems are observed to be helpful in treating indigestion, bronchitis, constipation, coughs and asthma. The roots are used as a contraceptive and are chewed by singers to improve their voice and prevent voice cracking (Pradhan et al., 2013). Historically, PB leaf was known to the communities for hundreds of years for its curative effects such as to reduce/save you foil breath and body odor, throat and lung problems, cough prevention and healing, to lessen unwanted vaginal secretion and to prevent itching as a result of fungus and internal/external bacteria. PB leaves are used for the remedy of numerous disorders and claimed to have detoxication, antioxidation and antimutation effect (Rai et al., 2010).

In Ayurveda, PB leaves are also prescribed to treat halitosis, bronchitis and elephantiasis. The Ayurvedic practitioners claim that the leaves are anthelminthic, aphrodisiac, carminative, laxative, stomachic and tonic. These leaves are used in the treatment of diseases, which are difficult to treat. In the Unani medicine, the PB leaves are regarded as styptic and a vulnerary and used to improve the appetite and taste, to strengthen teeth and as tonic for the brain, heart and liver. The leaf extract is also recommended in a mixture ingested to treat gonorrhea. This plant is used for the treatment of dysentery, fever, gastritis, rheumatism and leucorrhea. It is also used to eliminate body odor. PB leaves are used in lotions and paste, which are applied to nose ulcers, swellings and wounds. Leaf extracts are also used as eardrops and eye drops. The essential oil of PB leaves is used as an external application for treating catarrh and breast swellings. The leaf and root, mixed in oil, are used as an ointment to treat hard tumours. PB leaves have been found to have various pharmacological actions such as analgesic, antiallergic, anticancer, anticarcinogenic, antidiabetic, antiinflammatory, antiinfective, antileishmanial, antimutagenic, antioxidant and radio protective activities. They have also shown antiamoebic, antifungal, antimicrobial, antifertility, antiplatelet, cardiovascular, hepatoprotective, and immunomodulatory activities (Lei et al., 2003). Administration of PB extracts led to an increase in metabolic function in animals maintaining their body weight despite being on a high fat diet (Chen et al., 1995; Ghani et al., 2016).

1.10.1 Antibacterial Activity

The well known function of PB leaf is the suppression of halitosis. To find out the active phytochemicals against halitosis, methanol extract of fresh leaves was evaluated in vitro using plate and broth MIC assays, bio film assay, saliva chip model and a conductometric method. The results showed that the active constituent, allylpyrocatechol, is responsible for the antimicrobial activity against various obligate oral anerobes (Ramji etal., 2002). PB leaf also shows antimicrobial activity against *Streptococcus pyrogenes*, *Staphylococcus aureus*, *Proteus vulgaris*, *Escherichia coli* and *Pseudomonas aeruginosa*. The leaf extract also contains bactericidal activity against urinary tract pathogenic bacteria such as *Enterococcus faecalis*, *Citrobacter koseri*, *Citrobacter freundii* and *Klebsiella pneumoniae*. Dermatophytosis, a disease of the keratinized parts of the body (skin, hair and nail) caused by three genera of highly specialized fungi called the dermatophytes, is also cured by it. A study was undertaken to evaluate the protective and healing effects of allylpyrocatechol against the indomethacin induced stomach ulceration in rat model with encouraging results (Evans et al., 1984; Kaveti et al., 2011).

1.10.2 Antidiabetic Activity

The aqueous and ethanolic extracts of PB leaf possess significant hypoglycemic activity when tested in fasted normoglycemic rats. The extracts showed antihyperglycemic activity in the external glucose level using glucose tolerance test. PB leaf has the ability of lowering blood glucose levels of streptozotocin induced diabetic rats, which indicated that the leaf extracts also have insulin mimetic activity (Arambewela et al., 2005a).

1.10.3 Antifertility Activity

An orally effective male contraceptive agent was developed with many doses of the leaf stalk extracts of *P. betle* for use in male mice. The antifertility results showed absence of toxicity in all metabolically active tissues of mice and the contraceptive efficiency was underscored revocable fertility after removal of treatment (Adhikary et al., 1989; Sarkar et al., 2000).

1.10.4 Antiinflammatory and Antiallergic Response

Hydroxychavicol extracted from PB leaf shows significant suppression of tumor necrosis factor alpha expression in human neutrophils, which confirm its role as an antiinflammatory agent. The antiinflammatory and antiarthritic activities are attributable to the down regulation of generation of reactive oxygen species (Ganguly et al., 2007). Through the inhibition of production of allergic mediators, PB leaf appears to be a potentially new therapeutic agent for the control of allergic diseases. The effects of ethanolic extract of PB leaf on the production of histamine and granulocyte macrophage colony stimulating factor (GM-CSF) by murine bone marrow mast cells and on the secretion of exotoxin and IL-8 by the human lung epithelial cell line, BEAS-2B, were investigated in vitro. The extracts significantly decreased histamine and GM-CSF produced by an IgE mediated hypersensitive reaction. The result suggested that PB may offer a novel healing approach for the cure of allergic disease through inhibition of construction of allergic mediators (Ganguly et al., 2007; Wirotesangthong et al., 2008; Hajare et al., 2011). Allylpyrocatechol a constituent from *P. betle* showed antiinflammatory effect in LPS-induced macrophages mediated by suppression of iNOS and COX-2 via the NF-κB pathway (Sarkar et al., 2008).

1.10.5 Antimalarial Activity

Essential oil of PB provided better protection from mosquito bites by *Anopheles stephensi* and *Culex fatigans* compared to known mosquito repellent citronella oil. PB oil provided more than 4 h protection against *A. stephensi* and *C. fatigans* when applied at the rate of 20 μl/cm^2, whereas citronella oil provided only 2.2 and 2.6 h protection. Thus, mosquito repellent activity of PB was better and long lasting than citronella oil (Al-Adhroey et al., 2011).

1.10.6 Antioxidant Activity

Leaf aqueous extracts of three PB landraces displayed significant antioxidant activity when assessed by in vitro process such as 2,2-diphenyl-1-picrylhydrazyl radical scavenging, hydroxyl radical scavenging, prevention of lipid peroxidation and superoxide radical scavenging (Choudhary and Kale, 2002; Dasgupta and De, 2004).

1.10.7 Insecticidal Activity

Insecticidal activities of PB oil toward insect pests have been investigated. The insecticidal effect of PB oil extracted from the leaves evaluated toward the *Callosobruchus maculatus* F. (bean weevil), *Sitophilus zeamais* M. (corn weevil) and lesser *Rhyzopertha dominica* F. (grain borer) the use of stored (aged) grain assay. The performance of remedies was assessed with the aid of determining the intense toxicity on grown up insects and the volume of stopping or suppressing the production of progenies. Thus, the biologically active element of PB leaf oil may additionally possess ovicidal properties that inhibited the improvement of eggs of *Callosobruchus maculates* into larvae. The experiment discovered that PB leaf oil turned into a fecundity lowering agent to adult *S. zeamais* and *R. dominica*. Likewise, the oil's ovicidal effect cannot be discounted. It was observed that essential oil from PB leaves became a promising grain protectant (Gragasin et al., 2006; Mohottalage et al., 2007).

1.10.8 Immunomodulatory Activity

Crude methanol extract of PB demonstrated a mixed type 1 and type 2 cytokine responses, suggesting a significant immune modulatory property of this plant (Kanjwani et al., 2008; Singh et al., 2009).

1.10.9 Chemopreventive and Anticancer Activity

The PB leaf showed dose dependent suppression of dimethylbenzanthracene induced mutagenesis in *Salmonella* Typhimurium strain TA98 with metabolic activation (Amonkar et al., 1986; Paranjpe et al., 2013).

1.10.10 Cholinomimetic Effect

Aqueous and ethyl acetate of PB extracts were assessed for their cholinergic responses using guinea pig ileum. It was detected that the spasmogenic effect was more in aqueous extract than ethyl acetate extract. In rabbit, jejunum K^+-induced contraction was inhibited by both extracts, with observed blockade in calcium channel. Thus, leaves comprise cholinomimetic and possibly calcium channel antagonist components, which may yield the basis for many biochemical properties shown by this plant (Gilani et al., 2000).

1.10.11 Neuropharmacological Profile

Hydroalcoholic extract of PB leaves exhibited improvement in the discrimination index, potentiating the haloperidol induced catalepsy, reduction in basal, as well as amphetamine induced increased locomotors activity and delay in sodium nitrite–induced respiratory arrest. These results suggest possible facilitation of cholinergic transmission and inhibition of dopaminergic as well as noradrenergic transmission by the extract (Vyawahare and Bodhankar, 2007).

1.10.12 Platelet Inhibition Activity

Hydroxychavicol from PB leaf was evaluated for the inhibitory effect on platelet aggregation. The results showed that hydroxychavicol is a potent inhibitor for cyclooxygenase activity and reactive oxygen scavenger. It inhibits the platelet calcium signaling, thromboxane B2 aggregation and production. Hydroxychavicol is a potential therapeutic agent for prevention and treatment of atherosclerosis and other cardiovascular diseases through its antiinflammatory and antiplatelet effects, without effects on hemostatic functions (Chang et al., 2007).

1.10.13 Protective and Healing Activity

PB leaf showed promising protective and healing activity. The protective and healing effect of allylpyrocatechol was evaluated against indomethacin induced stomach ulceration in rat model (Bhattacharya et al., 2007; Prabu et al., 2012).

1.10.14 Radioprotective Activity

The radioprotective property of ethanolic extract of PB leaves was studied as an alternative low cost preventive medicine from synthetic radioprotectants, which are reported to be toxic. The ability of the PB leaf extract in prevention of gamma ray induced lipid peroxidation and DNA damage in rat liver mitochondria was measured and evaluated to establish the mechanism of its radioprotective action. The investigation discovered significant immunomodulatory and superior radical scavenging activities of PB leaf, which is supposed to be due to the presence of phenolic bioactive compounds such as allylpyrocatechol and chavibetol. It demonstrates that PB leaf has a great potential as it is a very cheap and easily accessible natural radioprotectant to the common people (Ghosh and Bhattacharya, 2005; Bhattacharya et al., 2005; Ferreres et al., 2014).

1.10.15 Antidermatophytic Activity

Antidermatophytic activity of PB cream was investigated in crude ethanolic extracts of leaves along with *Alpinia galangal* rhizomes (Zingiberaceae) and *Allium ascalonicum* bulbs (Liliaceae). The results showed the promising antifungal effect of PB extracts (Trakranrungsie et al., 2008; Sharma et al., 2011).

1.10.16 Antihypercholesterolemic Activity

PB have antihypercholesterolemic and antioxidative potential as reported in a study of the putative antihypercholesterolemic and antioxidative properties of ethanolic extract of PB and its active constituent, eugenol, was evaluated in experimental hypercholesterolemia induced in Wistar rats. Eugenol, an active constituent of the PB extract, showed antihypercholesterolemic and other activities (Rekha et al., 2014; Venkadeswaran et al., 2014). Recently published study has also confirmed the antihypercholesterolemic activity in PB (Abdul et al., 2019). this study has also shown that PB constituents are also excreted in urine implying their passage through blood to kidney.

1.10.17 Antinociceptive Activity

PB is reported to have antinociceptive activity. PB extract markedly reduced the licking time in early and late phases of the formalin test in a bell shaped dose response curve. In the formalin test, the pain in the early phase is caused due to the direct stimulation of the sensory nerve fibers by formalin, whereas the pain in the late phase is due to the inflammatory mediators, such as histamine, prostaglandin, serotonin and bradykinin. It is reported that nonsteroidal antiinflammatory drugs reduce both phases of the formalin test (Majumdar et al., 2003). The highest antinociceptive activity was observed with hot water extract (HWE) and cold ethanolic extract (CEE) of PB and the antinociceptive activity of CEE was higher than that of HWE (Arambewela et al., 2005b; Arambewela et al., 2011).

1.10.18 Gastroprotective Activity

The gastroprotective activity of HWE and CEE of PB leaves in rats was also evaluated. In this study, the highest dose of HWE did not cause significant inhibition in acidity (both total and free) or pH of gastric fluid. From the investigation, it was established that the gastroprotective effect of PB was not mediated via inhibition of acid secretion in the gastric mucosa but by increasing its mucus content (Arawwawala et al., 2014).

1.10.19 Antiasthmatic Effect

The effect of asthma can be reduced significantly by PB extract; however, the effect of PB on human asthma is not well known. But from the studies conducted by Misra et al., (2014), it was concluded that PB has the ability to reduce bronchial asthma in guinea pigs.

1.10.20 Effect on Thyroid Function Activity

On examination of the impacts of PB leaf extract for 15 days on the changes in thyroid hormone concentrations, lipid peroxidation (LPO), superoxide dismutase (SOD) and catalase (CAT) activity tried in male Swiss mice, it was inspected that feeding of betel leaf extract has shown a twofold activity, depending on the various dosages. Whereas the minimum dose diminishes thyroxine (T4) and increases serum triiodothyronine (T3) concentrations, invert impacts were found at two higher dosages. Higher amount expanded LPO with lowering in SOD and CAT effect. In this way, with the least dosage, activity was reversed. The discoveries proposed that PB can be both stimulatory and inhibitory to thyroid work, basically for T3 production and lipid peroxidation in male mice, depending on the total administered amount (Panda and Kar, 1998; Kumar et al., 2010a; Abrahim et al., 2012; Sengupta and Banik, 2013; Panda et al., 2018; Panda et al., 2019).

Metabolite Profiling of *Piper betle* Landraces by Direct Analysis in Real Time Mass Spectrometric Technique

2

2.1 PLANT MATERIAL AND CHEMICALS

Since the plant has very wide distribution, efforts were made to analyze as many landraces as could be procured. Locally available eight landraces, namely, (1) Bangla, (2) Desawari, (3) Deshi, (4) Jalesar Green, (5) Jalesar White, (6) Kalkatiya, (7) Mahoba and (8) Saufia were collected from the local market *Paan Dariba,* Lucknow, Uttar Pradesh (UP) and were studied for fingerprinting. Similarly, 21 *Piper betle* (PB) landraces (Bangla, Bangla Mahoba, Deshi Bangla, Desawari, Jalesar Green, Jalesar White, Kalkatiya, Mahoba and Sirugamani from Lucknow, UP; Ganzaam, Jagarnathi Green, Jagarnathi White, Maghi White, Maihar Deshi and Sanchi from Varanasi, UP; and Bangla Meetha, Maghi, Meetha Patta, Saufia and Sanchi from Kolkata, WB) were obtained from different markets and institutions in India and analyzed to assess their therapeutic potential based on mass fingerprints. Fully grown mature leaves from flowering PB plants were obtained from the research station of Indian Institute of Horticultural Research, Bengaluru, to identify the characteristic differences in male and female plants. In addition, 63 PB landraces were collected from local and different outstation markets of major cultivating regions throughout India (Table 2.1) to the effect of edaphic/ environmental factors on landraces.

Furthermore, 13 PB landraces, namely, Meetha Patta (West Bengal), Sanchi (West Bengal) Shirpurkata, Kapoori, Assam Paan, Nagpuri, Jalesar Green, Jagarnathi, Deshi, Mahoba, Saufia, Jalesar White and Bangladeshi were selected for quantitative study. The voucher specimens of these landraces, viz., Meetha Patta, Jagarnathi, Jalesar Green, Deshi, Kapoori, Jalesar White, Saufia, Bangladeshi and Mahoba (Voucher No. KRA 24476, KRA 24477, KRA 24478, KRA 24479, KRA 24480, KRA 24481, KRA 24482, KRA 24483 and KRA 24484, respectively) were deposited in the herbarium of the CSIR-Central Drug Research Institute, Lucknow, India. The voucher specimens of Shirpurkata, Sanchi (West Bengal) and Assam Paan (Voucher No. IIHRBV9, IIHR BV 24 and IIHR BV 45, respectively) were collected from Indian Institute of Horticulture, Bengaluru, Karnataka. HPLC (high performance liquid chromatography) grade ethanol (Merck Darmstadt, Germany) was used for extraction, and LC-MS (liquid chromatography–mass spectrometry) grade methanol (Sigma Aldrich St. Louis, MO, USA) was used for sample preparation.

2.2 EXTRACTION AND SAMPLE PREPARATION

Finely chopped leaves (50 g) were suspended in percolator with (500 mL) ethanol, sonicated for 30 min in an ultrasonic water bath at 30°C for proper mixing and subsequently allowed to stand at room temperature for 48 h. The percolate was collected, and the extraction process was repeated six times to ensure complete extraction. The combined extract was filtered through filter paper (Whatman No. 1) and concentrated on Buchi rotary evaporator (Rotavapor-R2, Flawil, Switzerland) under reduced pressure of 20–50 kPa at

TABLE 2.1   *Piper betle* landraces with their collection or procurement sites

S. NO.	STATE	DISTRICT	NAME OF P. BETLE LANDRACES	S. NO.	STATE	DISTRICT	NAME OF P. BETLE LANDRACES
1	Kerala	Ernakulam	Aluva	29	Uttar Pradesh	Lucknow	Jalesar White
2	Kerala	Ernakulam	Aluva Southern	30	Uttar Pradesh	Lucknow	Mahoba
3	Kerala	Ernakulam	Ampro Odekkali Local-1	31	Uttar Pradesh	Lucknow	Saufia
4	Kerala	Ernakulam	Ampro Odekkali Local Paan	32	Uttar Pradesh	Lucknow	Bangla Mahoba
5	Kerala	Ernakulam	Ampro Odekkali Local Medicinal Paan	33	Uttar Pradesh	Lucknow	Deshi Bangla
6	Kerala	Ernakulam	Gauthi	34	Uttar Pradesh	Lucknow	Kapoori Vellaikodi
7	Kerala	Ernakulam	Kalli	35	Uttar Pradesh	Lucknow	Sirugamani
8	Kerala	Ernakulam	Kollam big	36	Uttar Pradesh	Banaras	Sanchi
9	Kerala	Ernakulam	Kollam small	37	Uttar Pradesh	Banaras	Maghi White
10	Kerala	Ernakulam	Maghai	38	Uttar Pradesh	Banaras	Jagarnathi Green
11	Kerala	Ernakulam	Perumbawoor local	39	Uttar Pradesh	Banaras	Jagarnathi White
12	Kerala	Ernakulam	Pheyapala	40	Uttar Pradesh	Banaras	Maihar deshi
13	Kerala	Ernakulam	*Piper chavya*	41	Uttar Pradesh	Banaras	Ganzaam

(Continued)

TABLE 2.1 (Continued) *Piper betle* landraces with their collection or procurement sites

S. NO.	STATE	DISTRICT	NAME OF P. BETLE LANDRACES	S. NO.	STATE	DISTRICT	NAME OF P. BETLE LANDRACES
14	Kerala	Ernakulam	Salem local	42	West Bengal	Calcutta	Sanchi West
15	Kerala	Ernakulam	Thirupunithura local	43	West Bengal	Calcutta	Maghi West
16	Kerala	Ernakulam	Trivandrum local	44	West Bengal	Calcutta	Meetha Patta
17	Maharashtra	Pune	Calcutta Bangla	45	West Bengal	Calcutta	Bangla Meetha
18	Maharashtra	Pune	Gachi	46	Karnataka	Bengaluru	Hy Leaves
19	Maharashtra	Pune	Helisa	47	Karnataka	Bengaluru	Halisaha (s) Female
20	Maharashtra	Pune	Kapuri Chintalpudi	48	Karnataka	Bengaluru	Godi Bangla Female
21	Maharashtra	Pune	Khasi	49	Karnataka	Bengaluru	Calcutta Bangla
22	Maharashtra	Pune	Malvi	50	Karnataka	Bengaluru	Cari 6 Lateral Leaf
23	Maharashtra	Pune	Shirpurkata	59	Karnataka	Bengaluru	Calcutta Bangla Female Main Leaf
24	Uttar Pradesh	Lucknow	Bangla	60	Assam	Guwahati	Sonapuri
25	Uttar Pradesh	Lucknow	Calcuttiya	61	Assam	Guwahati	Baroi Kamakhya
26	Uttar Pradesh	Lucknow	Desawari	62	Assam	Guwahati	Khasi Assam
27	Uttar Pradesh	Lucknow	Deshi	63	Assam	Guwahati	Assam Paan
28	Uttar Pradesh	Lucknow	Jalesar Green				

45°C resulting in a dark green semisolid mass. A stock solution (1.0 mg/mL) of the PB leaf extract was prepared in methanol and filtered through a 0.22 µm polyvinylidene fluoride (PVDF) membrane MILLEX GV filter unit (Millex GV, PVDF, Merck Millipore, Darmstadt, Germany) and used for direct analysis in real time mass spectrometric (DART-MS) analysis using glass capillary. Stock solutions were stored at −20°C for further use.

2.3 PREPARATION OF SAMPLES FOR DIRECT ANALYSIS IN REAL TIME MASS SPECTROMETRIC ANALYSIS

For DART-MS measurements, intact fresh leaf was used. The leaf was washed with tap water followed by distilled water to remove any foreign matter. Washed surface dried leaf (no free water) was chopped into small pieces and positioned in the MS gap with the help of forceps for analysis.

2.4 ANALYSIS CONDITIONS

The mass spectrometer used was a JMS-T100LC (AccuTOF, atmospheric pressure ionization time-of-flight mass spectrometer, Jeol, Tokyo, Japan) fitted with a DART ion source. The mass spectrometer was operated in positive ion mode with a resolving power of 6000 (full width at half maximum). The orifice 1 potential was set to 28 V, resulting in minimal fragmentation. The ring lens and orifice 2 potentials were set to 13 and 5 V, respectively. Orifice 1 was set at 100°C. The RF ion guide potential was 300 V. The ion source was operated with helium gas flowing at approximately 4.0 L/min. The gas heater was set to 300°C. The potential on the discharge needle electrode of the source was set to 3000 V; electrode 1 was 100 V and the grid was at 250 V. Freshly cut pieces of PB leaf was positioned in the gap between the source and mass spectrometer for measurements. Data acquisition was from m/z 50 to 1000. The acquisition time for the analysis was 0.3 min. Exact mass calibration was accomplished by including a mass spectrum of neat polyethylene glycol (PEG) (1:1 mixture PEG 200 and PEG 600) in the data file. m-Nitrobenzyl alcohol was also used for calibration. The mass calibration was accurate to within ±0.002 u. Using Mass Center software, the elemental compositions were determined on selected peaks.

2.5 MULTIVARIATE STATISTICAL ANALYSIS

Principal component analysis (PCA) was performed by Minitab 14 statistical analysis software to differentiate among PB landraces. Scaling of the data matrix was carried out in order to adjust the mean and variance of each peak to make observations scale free common platform for the statistical analysis. This pretreatment provides a mean value of 0 and a variance of 1 to each mass. The interrelationship (similarity or dissimilarity) between 21 PB landraces was carried out on the basis of 18 ions estimated by DART-MS technique and analyzed by cluster analysis (CA) (hierarchical clustering; distance measure—single linkage and Euclidean distances; Linkage rule—Ward's method). Among the total 18 ions, the most active ions were identified by using factor analysis (FA), PCA and CA. The analysis was done on the basis of 3 known biologically active ions (m/z 151, m/z 165 and m/z 193) to see the therapeutic potential. The CA was done after standardizing the ions (mean=0, SD=1). Neighbor joining (NJ) tree for PB landraces was generated from the m/z peak data such that only the polymorphic peak was included in the analysis and binary state criterion (present/absent) was used for generating NJ tree by FREETREE software using the Jaccard coefficient and the NJ method. The identification of gender of unknown PB (Sirugamani, Calcutta Bangla, Helisa, Shirpurkata, Khasi and Malvi) from known male (Kapoori Chintalpudi) and female (Gachi) was done by k-means (standardized, nonhierarchical) clustering. CA was also done on the basis of four peaks at m/z 151, 193, 235 and 252 by k-means (standardized, nonhierarchical) clustering. The PCA was also carried out to differentiate among the PB landraces obtained from different geographical regions. All statistical analyses were performed on STATISTICA Windows version 7.0 (StatSoft, Inc., USA).

2.6 METHOD VALIDATION

Fifteen repeats of each sample were used to check the reproducibility of the optimized DART-MS method. The chemometric validation procedures include test set validation, where some samples from the total experiments were used for testing purposes. The statistical model was tested and validated by randomly dividing the repeats of the sample in a known and unknown group. The known sample was used to construct the model, and unknown was used to test and validate the model. A total of 60% experimental results were used to build the PCA model for discriminating the gender and geographical sample of the PB leaf and the remaining 40% experimental results were used for testing and validation of the PCA model so built.

2.7 OPTIMIZATION OF DIRECT ANALYSIS IN REAL TIME MASS SPECTROMETRIC CONDITIONS

DART-MS conditions were optimized to study phytochemicals in PB leaf extracts. To optimize the DART-MS instrumental parameters, different potential and temperature for orifice 1 and 2, helium gas flow rates (at 2.0, 3.0, 3.5, 4.0, 4.2, 4.5 L/min), electrode and grid voltage and different gas heater tem peratures at 100°C, 150°C, 200°C, 250°C, 300°C and 350°C were varied to determine the optimal ionization conditions for the leaf to obtain maximum and accurate peaks (*m/z*) with no fragmentation. Finally, optimized condition at 28 V potential of orifice 1, 13 V potential for ring lens and 5 V potential for orifice 2 and helium gas flow at 4.2 L/min with heater temperature of 300°C was set to record the spectra as described in the previous section, to obtained characteristic fingerprint spectra for the leaf. The acquisition time for fast analysis was optimized and 0.30 min was satisfactory for acquiring the sample spectra accurately. At low temperature (100°C), desorption and ionization of mainly low molecular weight compounds (below 200 Da) were observed in the positive ion mode, whereas in the negative ion mode, no peak above 150 Da was observed. At high temperature (350°C), the observed ion intensity of some of these compounds decreased markedly. A temperature of 300°C was found as the optimal temperature for ionization. However, in negative ionization mode, no remarkable improvement in ionization and intensity of peaks was observed and spectra were approximately same on both the temperatures 100°C or 300°C. Flow rates of helium gas were also observed to have an influence on the DART-TOF-MS fingerprint patterns. The number of metabolites detected increased with increased flow rate, but gas flow rates above 4.0 L/min led to depression of the signal intensity of the sample. In positive ion mode, various metabolites of higher molecular weight (approximately ≤700 Da) were detected and analyzed, whereas in the negative ion mode, only low molecular weight compounds below 200 Da were observed (Table 2.2). Positive ionization mode was selected for the analysis of intact PB leaf. Under these optimized conditions, the DART heated gas stream is used for direct analysis of the intact

TABLE 2.2   Optimization of DART-MS parameters for the analysis of *Piper betle* leaf in positive and negative ionization modes

S. NO.	PARAMETERS	OPTIMIZED SETTINGS	OPTIMIZED SETTINGS
1.	Ion mode	Positive	Negative
2.	Resolution (full width at half maxima)	6000	6000
3.	Orifice 1 potential	28 V	−28 V
4.	Orifice 2 potential	5 V	−5 V
5.	Orifice 1 temperature	100°C	100°C
6.	Ring lens potential	13 V	−15 V
7.	RF ion guide potential	300 V	−280 V
8.	Helium flow rate	4.0 L/min	4.2 L/min
9.	Needle voltage	3000 V	3000 V
10.	Discharge electrode	100 V	−150 V
11.	Grid electrode	250 V	−250 V
12.	Gas beam temperature	300°C	250°C
13.	Sampling time	20 sec	60 sec
14.	Mass range	50–1000 Da	100–500 Da
15.	Peak voltage	600 V	600 V
16.	Detector voltage	−2300V	+2500V
17.	Acquisition rate	3.0 sec	5.0 sec
18.	Sensitivity	High (maximum)	Low (minimum)

leaf and ethanolic leaf extract. The fingerprint patterns observed from intact leaf showed a maximum number of metabolites. Hence, intact leaf in positive ionization mode was used to generate fingerprint pattern of leaf; however, ethanolic extract was also used to calculate the high resolution mass spectrometric data for some compounds. To calculate the exact mass of the compounds, external calibration was used by recording a mass spectrum of neat PEG of 1:1 mixture with PEG 200, PEG 400 and PEG 600, which covers the mass range from 100 to 1000 Da. The PEG spectrum was mixed (added) with/in the data file of the sample and then using the Mass Center software, the exact mass and elemental composition was determined on selected *m/z* values.

The phytochemical screening or metabolite profiling of PB leaf was aimed at identification and dereplication of phytochemicals in intact leaf. Many secondary metabolites have been reported in PB leaves such as phenolics, flavonoids, phenylpropanoids, sesquiterpenes and sterols. Since phytochemicals present in PB leaf usually comprise C, H, O, and N elements producing characteristic molecular formulae and isotopic peak patterns, it is possible to identify the formula of the compounds by high resolution mass spectrometry (HRMS) with a high degree of confidence without using pure standards. A mass accuracy corresponding to 5-ppm mass error was considered to be accurate enough for high confidence formula confirmation/determination. HRMS, preferred for the screening of targeted as well as nontargeted phytochemical analysis, was employed for the determination of PB metabolite molecular formulae. Phytochemical analysis showed the presence of mainly terpenes and phenols in PB leaves. These constituents vary in the different PB landraces. The DART mass spectra did not show peaks attributable to terpenes except a small peak at m/z 205 corresponding to sesquiterpenes. However, peaks were observed at m/z values corresponding to the reported phenols and their acetates in PB. In the phytochemical screening, elemental compositions for ten compounds were detected based on exact mass measurements. The compounds at m/z 135, 151, 165, 177, 193, 207 and 235 were identified and corresponding to chavicol (CHV), allylpyrocatechol (APC), chavibetol, CHV acetate, APC acetate, chavibetol acetate, and APC diacetate, respectively. The elemental compositions are as follows: the peak at m/z 127.0397 $[M+H]^+$ corresponds to pyragallol $[C_6H_7O_3]$, at m/z 135.0809 $[M+H]^+$ corresponds to CHV $[C_9H_{11}O_1]$, at m/z 149.0978 $[M+H]^+$ corresponds to estragole $[C_{10}H_{13}O_1]$, at m/z 151.0766 $[M+H]^+$ corresponds to hydroxychavicol {4-APC} $[C_9H_{11}O_2]$, at m/z 165.0925 $[M+H]^+$ corresponds to chavibetol or eugenol $[C_{10}H_{13}O_2]$, at m/z 166.0868 $[M+H]^+$ corresponds to phenyl alanine $[C_9H_{12}N_1O_2]$, at m/z 177.0942 $[M+H]^+$ corresponds to CHV acetate $[C_{11}H_{13}O_2]$, at m/z 179.0734 $[M+H]^+$ corresponds to coniferaldehyde $[C_{10}H_{11}O_3]$, at m/z 193.0857 $[M+H]^+$ corresponds to APC acetate $[C_{11}H_{11}O_3]$, at m/z 205.1972 $[M+H]^+$ corresponds to β-caryophyllene $[C_{15}H_{25}]$, at m/z 207.1029 $[M+H]^+$ corresponds to chavibetol acetate $[C_{12}H_{15}O_3]$, and at m/z 209.0819 $[M+H]^+$ corresponds to sinapinaldehyde $[C_{11}H_{13}O_4]$; a very small and less prominent peak at m/z 223.0819 $[M+H]^+$ could correspond to nerolidol $[C_{15}H_{27}O_1]$; one of the most abundant peaks at m/z 234.0965 $[M+H]^+$ may correspond to APC diacetate $[C_{13}H_{15}O_4]$; and finally, one flavonoid peak at m/z 301.1081 was observed, which was previously reported from PB leaf and could correspond to hydroxy-5,7-dimethoxyflavanone $[C_{17}H_{17}O_5]$. Some peaks were also observed in abundance in the leaf, such as peaks at m/z 175.0759 $[C_{11}H_{11}O_2]$ and 252.1235 $[C_{13}H_{18}N_1O_4]$, which could not be identified. Several other peaks were remained unidentified due to lack of sufficient data and literature support. The exact mass values of selected compounds measured are given in Table 2.3. Since some of the phenols have the same molecular weight, it was not possible to distinguish those using DART mass spectra alone. The peak at m/z 151 could be due to APC or carvacrol. The peak at m/z 165 could be due to eugenol or chavibetol. Since both have the same molecular formula, a distinction could not be made.

TABLE 2.3   DART-MS based Identification of Phytochemicals in *Piper betle* leaf

S. NO.	MOLECULAR WEIGHT	MEASURED PEAK TYPE	MEASURED MASS	CALCULATED MASS	MOLECULAR FORMULA	ERROR Δ (MMU)	IDENTIFICATION
1.	134	[M+H]+	135.0812	135.0809	$C_9H_{11}O_1$	0.29	Chavicol
2.	148	[M+H]+	149.0966	149.0978	$C_{10}H_{13}O_1$	1.17	Estragole
3.	150	[M+H]+	151.0759	151.0766	$C_9H_{11}O_2$	0.70	Allylpyrocatechol
4.	164	[M+H]+	165.0915	165.0925	$C_{10}H_{13}O_2$	0.95	Chavibetol/eugenol
5.	165	[M+H]+	166.0886	166.0868	$C_9H_{12}N_1O_2$	1.84	Phenyl alanine
6.	174	[M+H]+	175.0776	175.0759	$C_{11}H_{11}O_2$	1.73	Unknown
7.	176	[M+H]+	177.0915	177.0942	$C_{11}H_{13}O_2$	2.64	Chavicol acetate
8.	192	[M+H]+	193.0864	193.0857	$C_{11}H_{11}O_3$	−0.77	Allylpyrocatechol acetate
9.	204	[M+H]+	205.1956	205.1972	$C_{15}H_{25}$	1.57	β-Caryophyllene
10.	206	[M+H]+	207.1021	207.1029	$C_{12}H_{15}O_3$	0.78	Chavibetol acetate
11.	234	[M+H]+	235.0970	235.0965	$C_{13}H_{15}O_4$	−0.53	Allylpyrocatechol diacetate
12.	251	[M+H]+	252.1228	252.1235	$C_{13}H_{18}N_1O_4$	−0.71	Unknown

Source: Reproduced from Bajpai et al., 2010 with permission from John Wiley & Sons.

2.8 DISCRIMINATION OF *PIPER BETLE* LANDRACES

A representative DART mass spectrum of PB leaf is given in Figures 2.1 and 2.2. Clear differences based on the landraces can be observed in the spectra as shown in Table 2.4. A peak at *m/z* 135 corresponding to CHV is seen only in the landrace Saufia. APC was absent only in the landrace Desawari, whereas chavibetol was not present in Desawari and Mahoba landraces. CHV acetate was present only in Bangla and Saufia landraces. The peaks at *m/z* 193, 207, 235 and 252 were common to all the landraces. DART mass fingerprint of landraces of Jalesar White and Jalesar Green showed a very minute difference in phytochemical profile (Table 2.4). It is not easy to identify and differentiate all the components independently based on their mass spectral data. MS based chemical profiling generates complex data sets, which need sophisticated software to interpret. Visualization is a key aspect as the data contained a number of variables.

PCA provides an informative first look at the data set structure and relationships between groups. The PCA groups of the samples rely solely on the

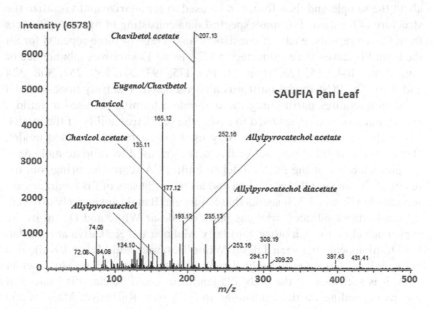

FIGURE 2.1   DART mass spectrum of Saufia.

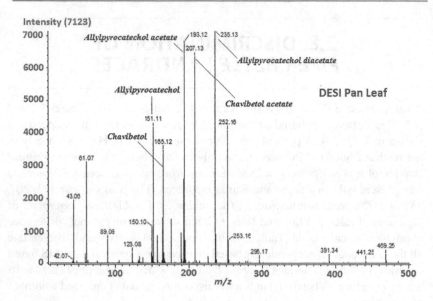

FIGURE 2.2   DART mass spectrum of Deshi.

information of the measured data and do not require any extra knowledge about the sample and therefore, can be used to summarize and visualize the structure of the data. The mass spectral data consisting of five sets (five sets from fifteen repeats, each set consists of an average of three repeats) for all the eight PB leaves were subjected to PCA using 14 variables (abundance of ions at m/z 104, 115, 123, 150, 151, 165, 175, 193, 207, 235, 252, 308, 324 and 352). The PCA model built was also validated. The truly honest model validation requires partitioning the data into a training set used to build a model and a validation set used to assess the predictive ability of the model, where the validation set is in no way used to generate the trained model. Hence, two sets (blue dot) out of five sets were used to validate and check the predictability of the PCA model so built. PCA score plot brings out the relationship among all the PB data and all the eight sets of PB landraces are separated in Figure 2.3. It seems that Saufia and Bangla are entirely different from all other landraces, whereas pairs of Jalesar White and Desawari are very much close to each other. Similarly, Mahoba and Kalkatiya are similar, but Deshi appears between Jalesar White/Desawari and Mahoba/Kalkatiya pairs. Jalesar Green is entirely different from Jalesar White. In a wider view, it is seen that all the eight PB landraces could be classified into four groups depending on their positions in PCA plot. Kalkatiya, Mahoba and Jalesar Green could be kept in one group, whereas Jalesar White, Desawari

TABLE 2.4 DART mass spectral data of *Piper betle* landraces, Bangla (1), Desawari (2), Deshi (3), J. Green (4), J. White (5), Kalkatiya (6), Mahoba (7), and Saufia (8)

	PIPER BETLE *LANDRACES*							
M/Z	*1*	*2*	*3*	*4*	*5*	*6*	*7*	*8*
104	a	x	a	a	x	x	a	a
118	x	x	a	a	x	a	a	a
132	x	x	a	a	x	a	a	a
135	a	a	a	a	a	a	a	x
151	x	a	x	x	x	x	x	x
163	x	a	a	x	x	x	x	x
165	x	a	x	x	x	x	a	x
166	a	x	a	a	x	a	a	a
175	x	a	x	x	x	x	x	a
177	x	a	a	a	a	a	a	x
193	x	x	x	x	x	x	x	x
205	x	a	a	x	x	x	a	x
207	x	x	x	x	x	x	x	x
235	x	x	x	x	x	x	x	x
252	x	x	x	x	x	x	x	x

a, absent; x, present.
Source: Reproduced from Bajpai et al., 2010 with permission from John Wiley & Sons.

and Deshi Paan could be in another group and the Bangla and Saufia are completely separated from all the remaining landraces and form another group. It is evident from this study that PCA could effectively segregate the landraces based on their characteristic chemical signatures, underscoring the effectiveness of this method in identifying the uniqueness of a landrace group.

2.9 DIFFERENCES IN *PIPER BETLE* MALE AND FEMALE LEAF METABOLITE PROFILE

Though human awareness of dioecy is as old as Babylonian times (ca. 2300 BCE) when different sexes were known for date palms, its significance in terms of metabolites has escaped the attention in the past. In plants, gender

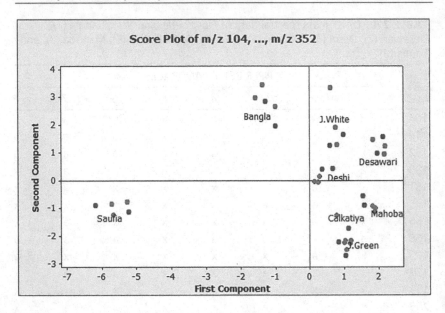

FIGURE 2.3 Validated PCA score plot of *Piper betle* landraces. (Reproduced from Bajpai et al., 2010 with permission from John Wiley & Sons.)

distinctions are generally based on the sex of the flowers, and the dioecy is cryptic than explicit as in case of animals. Reports on gender based differences in some physiological attributes and at the metabolite levels have been reported (Kumar et al., 2006; Tripathi et al., 2006). Thus, the bio/chemodiversity due to gender is over and above the diversity observed within the species due to location and developmental stages, etc. Thus, in nature, it is possible to have a gender associated qualitative or quantitative difference that may ultimately affect the level of metabolites and the efficacy of that taxon.

In PB, gender based differences in biological activities were reported by Tripathi et al., (2006). Higher amounts of phenols and antioxidants in the female plants are the part of putative defense function against herbivore attack. Knowing the gender of dioecious plant at vegetative stage is important when seed or fruit is the commercial product, as it will help in reducing the overall cost by elimination of male plants. It is also important in case of dioecious medicinal plants where only vegetative parts are being used, as in PB where gender based differences were reported in biological activities (Tripathi et al., 2006). In PB, molecular tools such as randomly amplified polymorphic DNA (RAPD) were also used for biodiversity analysis. CA based on RAPD data showed three major groups in PB: Bangla, Kapoori, and others. Though this is one of the quickest methods for molecular analysis of population, it involves sample preparation and other procedures before the results are obtained. DART-MS is now widely employed for population screening of economically important dioecious plants where vegetative parts (only or largely) are being used for optimum utilization of resources. From the DART mass spectra of PB leaves, it is evident that there are differences in the peaks between male and female landraces. The most abundant peak in the mass spectra of the leaves of all male and female landraces except Sirugamani female is the peak at m/z 252 [$C_{13}H_{18}N_1O_4$], whereas the peak at m/z 195 corresponding to methoxy eugenol is abundant in all female leaves except Khasi paan. However, the peak at m/z 301 due to hydroxy-5,7-dimethoxyflavanone is only observed in the male landraces. The relative abundance of APC was high in female leaf. APC acetate and APC diacetate were observed marginally higher in female landraces. The peaks at m/z 150, 151, 164, 165, 175 193, 207, 235, and 252 are common to all male and female landraces. Other major peaks seen in the spectra at m/z 136, 137, 149, 177, 189, 224, 236, 295, 300, 301, 308, 352, 366, 431, 469, and/or 486 are specific to individual landraces. Peak at m/z 157 was specific only to Sirugamani and Gachi landraces, whereas peak at m/z 366 was observed only in Helisahar leaf.

In the DART-MS of male and female landraces, peaks at m/z 151 (APC), 193 (APC acetate), 235 (APC diacetate), and 252 were more abundant. The peak at m/z 252 was the most abundant in male landraces ($\geq$36.00%). From the DART-MS of PB landraces, the normalized relative abundances of 27 ions (m/z) were extracted. Peaks that showed differences based on percentage abundance between the two sexes were used for CA, and out of total 27 ions, 4 ions, viz., m/z 151, m/z 193, m/z 235, and m/z 252 were found to be discriminating ions, which classified the sex of unknown PB 100% correctly using K-means clustering analysis. In this analysis there were eight PB landraces with known sex. Sex of two PB landraces (Tellaku Chintalapudi male and Gachi female) was used as known (disclosed) to create a statistical model, whereas sex of remaining six landraces was kept undisclosed (unknown) to analyze the result (Table 2.5). Cross validation was used to validate the analysis for grouping of landraces into male and female (Shirpurkata male and Helisahar female) with the known (disclosed) male and female while treating other landraces as unknown (undisclosed) male and female landraces. The results obtained from these experiments were quite similar, which were further confirmed by hierarchical clustering (Figure 2.4). On the basis of peak abundance in PB leaf, the identification of sex (gender) of unknown PB (Sirugamil, Calcutta Bangla, Helisa, Shirpurkata, Khasi, and Malvi) from

TABLE 2.5   Gender determination of unknown *Piper betle* on the basis of peak abundance

KNOWN SEX		UNKNOWN SEX	
P. betle	SEX	*P. betle*	SEX
Tellaku Chintalpudi	Male	Shirpurkata	Male
Gachi	Female	Calcutta Bangla	Female
		Helisahar	Female
		Malvi	Female
		Khasi	Female
		Sirugamani	Female

Source: Reproduced from Bajpai et al., 2012a with permission from Indian Academy of Sciences.

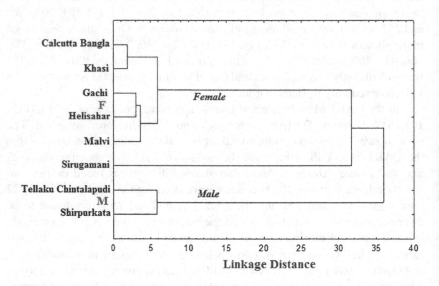

FIGURE 2.4   Identification and validation of gender in *Piper betle* using PCA. (Reproduced from Bajpai et al., 2012a with permission from Indian Academy of Sciences.)

known male (Kapoori Chintalpudi) and female (Gachi) was successfully done. Hierarchical clustering showed two different clusters representing male and female landraces. Using this developed and validated hierarchical clustering approach for identification of sex of unknown PB landraces, 38 landraces were differentiated in male and female group as shown in Figure 2.5.

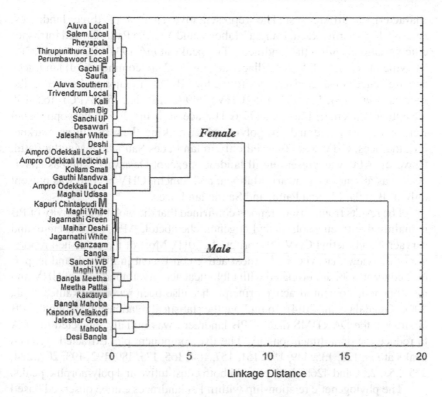

FIGURE 2.5   Identification of gender of thirty eight *Piper betle* landraces.

2.10 PHYTOCHEMICAL PROFILING TO CLASSIFY THE THERAPEUTIC POTENTIAL OF *PIPER BETLE* LANDRACES

PB leaves are used as a folk medicine, and several of its medicinal properties have been recently validated (Toprani and Patel, 2013; Fazal et al., 2014). The PB phenols and corresponding acetate/diacetate are mainly responsible for the biological activities of the leaves. These constituents vary in the different PB landraces. It is estimated that in India, nearly 150 PB landraces are documented, and in order to find out the suitable landraces for medicinal use, it is important to have their chemical profile. 21 PB landraces collected from different locations in India were analyzed and evaluated based on the molecules detected in leaves. DART-MS profiles of 21 PB landraces showed characteristic differences. The representative spectra of three landraces, namely, Jagarnathi Green, Bangla Mahoba, and Meetha Patta showed intraspecific variations within the landraces. The peaks at *m/z* 193 (APC acetate), 207 (chavibetol acetate), 235 (APC diacetate), and 252 are common to all landraces and are hence considered as constitutive for PB. Other major peaks seen in the spectra at *m/z* 104, 118, 132, 135 (CHV), 149 (estragole), 151 (APC), 163, 165 (chavibetol/eugenol), 166, 175, 177 (CHV acetate), and 205 (β-caryophyllene) were not constitutive and are polymorphic, making them useful as markers for landraces. CHV was observed only in landraces Saufia and Meetha Patta. However, APC was present in all landraces except Desawari, whereas chavibetol was absent in Desawari, Mahoba, and Sanchi. CHV acetate was present only in Bangla, Meetha Patta, and Saufia landraces.

The results from various reports confirmed that the biological activity of PB is mainly due to eugenol, methyl eugenol, chevibetol, APC, APC acetate, and carvacrol and methyl CHV (Nagori et al., 2011). Most of the published reports are of the view that APC is the most active constituent in PB leaf, and its percent concentration is correlated with biological activity. Importance of CHV and total phenolic content as active principle has also been reported (Rathee et al., 2006; Gundala et al., 2014). To find out the statistical relationship among the PB landraces, the DART-MS data of PB landraces was initially subjected to PCA to reduce the data dimensionality. The five component plot extracted eighteen peaks at *m/z* 123, 135, 149, 150, 151, 157, 164, 165, 175, 189, 192, 193, 207, 224, 235, 252, 259, and 426, which contain both constitutive and polymorphic peaks.

The phylogenetic relationship within PB landraces can be observed based on binary criteria (presence/absence) of peaks in the mass spectra. Phylogenetic relationship between PB landraces was evaluated by NJ tree using only polymorphic peaks. Constitutive peaks present in all the PB landraces tested were not included in this analysis. Based on NJ tree, the landraces were grouped into two broad clusters. The first cluster includes the landraces PB2 (Bangla Meetha), PB3 (Bangla Mahoba), PB4 (Deshi Bangla), PB6 (Deshi), PB8 (Jalesar Green), PB10 (Jagarnathi Green), PB13 (Maghi White), PB14 (Maghi), PB15 (Mahoba), PB18 (Sanchi V), PB19 (Sanchi WB), and PB21 (Sirugamani), whereas the second cluster consisted of PB1 (Bangla), PB5 (Desawari), PB7 (Ganzaam), PB9 (Jalesar White), PB11 (Jagarnathi White), PB12 (Kalkatiya), PB16 (Maihar Deshi), PB17 (Meetha Patta), and PB20 (Saufia). Such a grouping of the landraces based on polymorphic peaks also suggests that the secondary metabolite profiles of these landraces may include a set of PB specific metabolites along with two different groups of the peaks.

A scrutiny of the distance matrix of pairs of landraces indicates that landrace pairs PB3-PB14, PB4-PB6, PB4-PB21, and PB6-PB21 have the least distance value (maximum similarity), whereas the greatest distance values are for the pairs PB5 with PB4, PB6, PB8, PB10, PB13, PB15, and PB18–21, indicating that these landrace pairs are the most dissimilar in their DART-MS profiles. It is because of these trends that the landraces are divided into the two broad clusters. Percent ionization of bioactive molecules is also taken as relative parameter to compare the biological activity of landraces on this assumption that ions are directly related with molecules and ionization is done in natural form; hence, percent ionization of ions can be compared. The order of activity for PB landraces could be Sirugamani > Deshi Bangla > Kalkatiya > Deshi > Jalesar Green based on total phenolic content.

Most of the earlier findings indicate that biological activity is due to APC. It is evident from the result that Mahoba > Sirugamani > Deshi Bangla > Jalesar Green are the best among the 21 landraces analyzed. If we evaluate on the basis of CHV, Saufia > Bangla Mahoba > Deshi Bangla > Deshi > Meetha Patta (WB) are the most potential landraces for drug development. The clustering of PB landraces based on FA/PCA seems to indicate the potential biological activity. Findings of this study could be useful in screening the landraces for their potential medicinal uses. All our predictions discussed here are based on these findings.

Percent fractions of eighteen ions from 21 PB landraces were estimated by averaging of relative abundance of ions observed in DART-MS. Three biologically active ions at m/z 151, m/z 165, and m/z 193 were detected in almost all the PB landraces. The clustering of 21 PB landraces was further done on the basis of these three ions that are summarized in Figure 2.6. The average of three ions clustered 21 landraces into two broad groups. Landraces Bangla, Bangla Mahoba, Deshi Bangla, Deshi, Jalesar Green, Jalesar White, Kalkatiya, Mahoba, and Sirugamani with higher APC contents cluster in one group and are considered to have a high therapeutic potential.

Landraces Bangla Meetha, Desawari, Ganzaam Udisa, Jagarnathi Green, Jagarnathi White, Maghi white, Maghi, Maihar Deshi, Meetha Patta, Sanchi V, Sanchi WB, and Saufia with low APC or higher CHV contents cluster in other group and are considered to have a low therapeutic potential. The constituents of these two broad groups are based on three ions, though slightly different than in the phylogenetic tree; however, there was a broad similarity between the two trees to support the contention that the DART-MS profile for APC and CHV is a good predictor of the relative content, which can be correlated to their therapeutic potential. It is comparable with the earlier reports on the biological activity of these molecules. But there are few reports based on landraces, and all biological activities reported are from crude extract of PB prepared from leaves bought from local market without any rigorous characterization. These findings are indicative and may require further validation by quantitation and testing the biological activities as earlier reported by Tripathi et al., (2006) based on the level of total phenols in PB landraces. The quantitation of

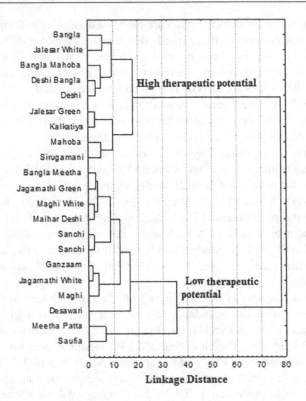

FIGURE 2.6 Tree view of therapeutic potential cluster for twenty one *Piper betle* landraces. (Reproduced from Bajpai et al., 2012b with permission from SAGE Publications Inc.)

bioactive compounds such as APC diacetate, eugenyl acetate, and eugenol and antimicrobial activity were evaluated in different landraces was also accomplished by Pandey et al., (2014), and the results validate the therapeutic potential as predicted in Figure 2.6.

2.11 DISCRIMINATION OF THE *PIPER BETLE* LANDRACES FROM DIFFERENT GEOGRAPHICAL ORIGIN

PB is cultivated throughout India except the dry northwestern parts. It grows best under the shaded, tropical forest ecological conditions with a rainfall of about 2250–4750 mm and relative humidity and temperature ranging from 40%–80% and 15°C–40°C, respectively. The cultivation of PB in India provides an example of growing from very mild climatic region to areas where climatic changes are sharp. The cultivation of PB in subtropics is the first example of anthropogenic transfer of a plant from its natural habitat to adverse condition where it just cannot grow on its own. In India, the favorable climate for naturally growing of PB plant is available in the state of West Bengal (WB) along with some places from Assam (As), Kerala (KL) and Karnatka (KA). Some secondary metabolites of this plant have important ecological functions such as resistance against diseases and herbivores.

The biosynthesis of secondary metabolites also depends on the environmental factors such as annual rainfall, humidity, precipitation, temperature, soil, and length of the vegetation period. These environmental factors affect the concentration of particular compounds, which is required for survival of the plants. Hence, the concentrations of compounds varied with change of environmental condition and locations and also with herbivores and epiphytes. Same plants growing in climatic variation are likely to exhibit geographic variation in response to environmental conditions because of adaptive evolution. The change of phytoconstituents in plants cannot be distinguished based on morphological features or traditional methods. It is, therefore, important to develop efficient methods that will allow the discrimination on chemical basis among the geographically available samples for quality control purposes. There are more than 150 landraces of PB in India of which maximum are found in West Bengal. Initially, we have observed difference in PB landraces from the same location or in gender of landraces as discussed earlier to identify the variation in phytochemicals. It is evident from many studies that phytochemical variation may also occur due to geographical changes. To find out the geographical variation in PB leaf, first we generated DART mass spectra of different landraces collected from natural climatic zone and adopted zone of PB cultivation. The characteristic compounds can be observed in all six spectra. The compounds observed in the mass range from 150 to 450 Da are characteristic for all PB leaves, as qualitatively it is common to all landraces; hence, it may be fingerprint region for PB leaf using DART-MS. However, differences in relative abundances of these peaks were detected among the landraces. In PB fingerprint of all the landraces, peaks at m/z at 135, 151, 165, 177, 193, 207, and 235 corresponded to CHV, APC, chavibetol, CHV acetate, APC acetate, chavibetol acetate, and APC diacetate, respectively. The major qualitative differences in all the landraces were observed in the range m/z 160–500, which could be landrace specific fingerprint region. Differences in the abundances of these peaks were also observed. The maximum difference in phytochemical profile of PB leaf among all regions was observed particularly in the landraces of UP region where a maximum number of peaks were observed in the range m/z at 160–500 in the DART mass spectra (landrace specific fingerprint region).

DART mass spectra of PB landraces of six geographical regions were recoded; all the selected landraces were analyzed 15 times (15 repeats for each sample). Mass spectrometric data combined with chemometric analyses have been widely applied to classify samples of biological or geographical origin (Héberger, 2008). It is therefore necessary to resort to chemometric methods or multivariate techniques to visualize the geographical difference in the landraces. In the concluding analysis of the leaf data from different geographical locations, PCA was performed for dimensionality reduction in an attempt to distinguish characteristic profiles from the DART-MS data and to investigate any potentially existing grouping based on the geographical regions. From the analysis of all 63 landraces, 76 peaks (m/z) were extracted from DART-MS spectra as described in experimental section data processing. The relative abundance of these 76 peaks from all landraces was subjected to PCA. The PC1 vs. PC2 plot shows that the first two principal components are able to explain highest variance of almost 71.91% information contained in 23 peaks at m/z 123, 136, 149, 151, 175, 177, 193, 165, 205, 207, 224, 235, 252, 324, 338, 352, 366, 398, 400, 409, 431, 441, and 486 from which all landraces of six geographical regions were clearly discriminated with the best projection by these peaks. The PCA analysis showed that the landraces collected from UP and MH are closer to WB and can be grouped in one, whereas PB from KN and KL are closer to each other and can be in another group. The PB collected from Assam region was entirely different from all others. It can be observed from the PCA analysis that PB mostly found in UP and MH has maximum profile similarity with PB from WB and it might have been introduced from WB region. Similarly, the phytochemical profile of PB of KA is more analogous to KL and may have a common pool between them. It is, therefore, clear that DART-MS followed by PCA is an appropriate method for the clear differentiation of different PB landraces according to geographical regions.

LC-MS Analysis of *Piper betle* Leaf and Evaluation of In Vitro Antimicrobial Activity

3

3.1 PLANT MATERIAL, REAGENTS AND CHEMICALS

Thirteen landraces of *P. betle* were selected on the basis of their easy availability, namely, Meetha Patta (West Bengal), Sanchi (West Bengal), Shirpurkata, Kapoori, Assam Paan, Nagpuri, Jalesar Green, Jagarnathi, Deshi, Mahoba, Saufia, Jalesar White and Bangladeshi for quantitative study.

Acetonitrile (liquid chromatography–mass spectrometry [LC-MS] grade) and formic acid (analytical grade) were procured from Sigma-Aldrich, and water was purified using a Direct-Q 8 UV water purification system (EMD Millipore Corporation, Billerica, MA, USA). Ethanol (ACS, ISO, Reag. Ph Eur grade) was procured from Merck Millipore (Darmstadt, Germany). Analytical standards (purity ≥ 95%) of allylpyrocatechol-3,4-diacetate and eugenyl acetate were from Sigma Aldrich (St. Louis, USA) and eugenol and daidzein were from Extrasynthese (Z.I Lyon Nord, Genay Cedex, France).

3.2 EXTRACTION AND SAMPLE PREPARATION

The shade dried leaves of 13 landraces of *P. betle* were crushed and suspended in ethanol and the suspension was placed in an ultrasonic bath for 30 min and then left for 24 h at room temperature for extraction of secondary metabolites. The process of extraction was repeated three times by adding ethanol to the residue. The supernatant was filtered (using Whatman filter paper) to eliminate the molecular impurities and evaporated to dryness under reduced pressure (20–50 kPa) using a rotary evaporator (Buchi Rotavapor-R2, Flawil, Switzerland) at 40°C. Dried residues (1 mg) were weighed accurately, dissolved in 1 mL of acetonitrile and sonicated using an ultrasonicator (Bandelin SONOREX, Berlin, Germany). The solutions were filtered through 0.22 μm syringe filter (Millex GV, PVDF, Merck Millipore, Darmstadt, Germany) and diluted with acetonitrile to final working concentration for ultrahigh performance liquid chromatography–electrospray ionization–tandem mass spectrometric (UHPLC-ESI-MS/MS) analysis.

3.3 PREPARATION OF STANDARD SOLUTION

Standard stock solutions of 1 mg/mL of selected analytes (allylpyrocatechol-3, 4-diacetate, eugenyl acetate and eugenol) and internal standard daidzein were prepared in acetonitrile. Aliquot of each stock solution was mixed and diluted with acetonitrile to make the standard mixture and it was further diluted to provide a series of concentrations in the range of 1–500 ng/mL that were used for plotting calibration curve. The standard stock and working solutions were all stored at −20°C until use and vortexed prior to injection.

3.4 INSTRUMENTATION AND ANALYTICAL CONDITIONS

The UHPLC-ESI-MS/MS analysis was performed on a Waters Acquity UPLC™ system (Waters, Milford, MA, USA) interfaced with a hybrid linear ion trap triple quadrupole mass spectrometer (API 4000 QTRAP™ MS/MS system from AB Sciex, Concord, ON, Canada) equipped with an electrospray (Turbo V) ion source. The Waters Acquity UPLC™ system was equipped with a binary solvent manager, sample manager, column oven and photodiode array detector (PDA). AB Sciex Analyst software version 1.5.1 was used to control the LC–MS/MS system and for data acquisition and processing.

The chromatographic separation of phytoconstituents and analytical standards was achieved on an Acquity UHPLC BEH C_{18} column (50 mm×2.1 mm id, 1.7 µm) at a column temperature of 40°C. The analysis was completed with gradient elution of 0.1% formic acid in water (A) and acetonitrile (B) as the mobile phase. The 4 min UHPLC gradient system was as follows: 0–1.5 min, 30–70% B; 1.5–3 min, 70–70% B; 3–3.5 min, 70–30% B; 3.5–4 min, 30–30% B. Sharp and symmetrical peaks were obtained at a flow rate of 0.3 mL/min with a sample injection volume of 1 µl. The UV spectra (PDA) were recorded in the range of 190–400 nm and 254 nm was chosen for determination.

The identification and characterization of phytoconstituents were performed in positive ESI mode using precursor ion scan (Q1MS) and product ion scan (MS/MS) at unit resolution for both Q1 and Q3. Mass spectra were recorded by scanning the mass range *m/z* 100–1000 at a cycle time of 9 sec with a step size of 0.1 Da. Nitrogen was used as the nebulizer, heater and curtain gas as well as the collision activated dissociation (CAD) gas. The source parameters were ion spray voltage set at 5500 V, curtain gas, nebulizer gas (GS1) and heater gas (GS2) set at 20, 50 and 50 psi, respectively and source temperature set at 550°C. Compound dependent parameters, namely, declustering potential and entrance potential were set at 80 and 12 V respectively. The CAD gas was set at medium and the interface heater was on. The quantitative analysis was performed using multiple reactions monitoring (MRM) acquisition mode.

3.5 OPTIMIZATION OF ULTRAHIGH PERFORMANCE LIQUID CHROMATOGRAPHY–TANDEM MASS SPECTROMETRIC CONDITIONS

In order to achieve optimum separation in a short analysis time, the chromatographic conditions such as the column, mobile phase and gradient program were optimized. Two types of analytical columns, the Acquity BEH C_{18} column (50 mm×2.1 mm id, 1.7 μm) and the Acquity CSH C_{18} (100 mm×2.1 mm id, 1.7 μm), were compared. The results showed that the Acquity BEH C_{18} column (50 mm×2.1 mm, 1.7 μm) produced chromatograms with better resolution within a shorter time. Various combinations of mobile phases (water–methanol, 0.1% formic acid in water–methanol, water–acetonitrile and 0.1% formic acid in water–acetonitrile) at different flow rates (0.1, 0.2, 0.25, 0.3 and 0.4) and column temperatures (25°C, 30°C, 40°C and 50°C) were optimized for better chromatographic behavior and appropriate ionization. A good chromatographic separation was achieved within 4 min using gradient elution with 0.1% formic acid in water and acetonitrile at 40°C column temperature with a flow rate of 0.3 mL/min. The detection wavelength was set at 254 nm as most of constituents show absorbance at this UV wavelength. MS analyses were tested in both positive and negative ionization modes. The positive ionization mode was preferred due to high sensitivity. The most abundant pseudomolecular ion in Q1MS scan mode was selected as the precursor ion for product ion scan and the product ions were recorded at various collision energies (10–45 eV). The efficiency of selected parameters has been experimentally evaluated. Finally, the optimized UHPLC-MS/MS method was applied for quantitation of phenolics using daidzein as an internal standard.

3.6 QUALITATIVE ANALYSIS

Nineteen phytoconstituents including propenyl phenols and their derivatives, flavonoids, lignans, hydroxycinnamic acid derivatives, polyphenols and terpenes were tentatively identified and characterized. Among them, three propenyl phenols (allylpyrocatechol-3, 4-diacetate, eugenyl acetate and eugenol) were unambiguously identified based on their retention times, MS and MS/MS spectra compared with the authentic standards. The tentative identification of the other constituents was based on interpretation of their MS/MS fragmentation patterns and by comparison with the previously reported literature. The [M+H]+ ions, their MS/MS fragment ions and UV absorption maxima for compounds identified from leaf extract of *P. betle* are listed in Table 3.1.

TABLE 3.1 The [M+H]+ ions, MS/MS fragment ions and UV absorption maxima for compounds identified from leaf extracts of *Piper betle* by UHPLC-ESI-MS/MS experiment

SR. NO.	T_R (MIN)	PRECURSOR ION [M+H]+	MS/MS FRAGMENT IONS, M/Z (% ABUNDANCE)	IDENTIFICATION
1.	1.09	223	205(3), 195(1.3), 191(4.7), 181(100), 177(1.7), 163(4.7), 154(50), 149(23.4), 140(3), 121(1.7), 117(0.8), 103(2.1)	Nerolidol
2.	1.09	303	285(8.3), 257(20.3), 239(2), 229(48.8), 211(1.4), 201(23.5), 183(7.4), 165(24), 153(68), 137(100), 135(13.7), 121(15.2), 109(17), 93(3.3), 81(4),69(14.5), 65(1.5)	Quercetin
3.	1.21	177	162(2), 149(20), 134(12.3), 121(18), 117(100), 105(13.3), 91(47), 81(2.8)	Chavicol acetate
4.	1.5	291	273(1.6), 241(1.7), 227(3.1), 213(3.2), 207(11.1), 199(9.6), 165(11.5), 161(67.4), 147(15.6), 139(100), 123(25), 105(2.2), 95(1.2), 71(0.9), 57(0.2)	Catechin
5.	1.65	165	150(21), 137(65.6), 132(7), 133.1(39.2), 124(100), 122(5.6), 109(20.5), 105.1(96.2), 103(5.3), 91(5.3), 79(4.9)	Eugenol
6.	1.65	301	283(4.1), 269(100), 255(16.3), 241(83), 227(10), 209(20.1), 181(37.2), 163(15), 149(2.6), 107(3.5)	8-Hydroxy-5,7-dimethoxytlavanone

(Continued)

TABLE 3.1 (*Continued*) The [M+H]+ ions, MS/MS fragment ions, and UV absorption maxima for compounds identified from leaf extracts of *Piper betle* by UHPLC-ESI-MS/MS experiment

SR. NO.	T_R (MIN)	PRECURSOR ION [M+H]+	MS/MS FRAGMENT IONS, M/Z (% ABUNDANCE)	IDENTIFICATION
7.	1.73	151	133(10.3), 123(100), 110(27.1), 105(43.9), 103(5.9), 95(6.3), 91(5.7), 79(5.3), 77(4.5)	Hydroxychavicol
8.	1.73	235	225(9.5), 207(54.4), 205.3(17), 204.5(6.8), 198.8(6.8), 193.2(100), 174.8(19.7), 157.3(12.2), 150.8(58.5).	Allylpyrocatechol-3,4-diacetate
9.	1.73	357	325(4.4), 293(2.6), 273(7), 265(10.6), 233(7), 205(100), 191(15), 177(35.3), 163(3.5), 149(6.1), 123(99.1), 105(26.5), 86(1.7), 69(2.6)	Pluviatilol
10.	1.81	135	120(1.5), 107(100), 105(15.2), 103(45), 91(55.5), 77(48.5), 67(0.6), 59(0.6), 55(1.5)	Chavicol
11.	1.81	207	189.1(100),179.2(13.7), 174.1(48), 165.1(100), 161.1(11.1), 157(55.8), 150.1(10.7), 137(81.3), 130.9(14.7), 133(47), 122.8(8.8), 124.1(36.2), 105(30.3), 95(7.8), 83(5.8)	Eugenyl acetate
12.	1.93	127	109.4(50), 100.8(50), 98.7(100), 85.9(75), 81.3(75), 78.9(37.5), 69.3(37.5), 67.4(25), 66.6(25), 57.9(50).	Pyrogallol
13.	1.93	209	191(2), 181(4.5), 177(9.4), 167(6.5), 163(24.6), 149(1.2), 137(17.6), 135(100), 121(25.5), 123(17.2), 117(44.8), 107(32.5), 105(9), 93(6.9), 87(12.7), 83(3.2), 69(2.4)	Sinapinaldehyde

(Continued)

TABLE 3.1 (Continued)   The [M+H]+ ions, MS/MS fragment ions, and UV
absorption maxima for compounds identified from leaf extracts of *Piper betle* by
UHPLC-ESI-MS/MS experiment

SR. NO.	T_R (MIN)	PRECURSOR ION [M+H]+	MS/MS FRAGMENT IONS, M/Z (% ABUNDANCE)	IDENTIFICATION
14.	1.93	355	295(0.5), 263(18), 231(3.3), 203(24.4), 175(5), 163(100), 145(22.4), 135(14.5), 97(8.3), 85(3.3), 57(5)	Chlorogenic acid
15.	2.13	285	267(1.7), 243(0.3), 207(18.5), 193(88.8), 165(100), 151(30.5), 137(415), 133(7.3), 123(22.6), 107(1.1), 95(7.1), 86(0.8), 67(0.8)	3-(2,4,5-Trimethoxyphenyl)-2-acetoxyl-hydroxypropane
16.	2.13	193	175(13.4), 161.1(72.5), 156.9(10), 151.1(87), 133(18.7), 129(6.2), 123(100), 105(22.5)	Allylpyrocatechol monoacetate
17.	2.29	195	180(2.5), 177(6.1), 163(2.5), 153(39.4), 145(1.5), 135(100), 133(3), 129(3), 125(62), 123(16.4), 121(90.7), 117(14.8), 107(49.7), 105(39.4), 95(2), 93(26.1), 79(3)	Methoxyeugenol
18.	2.34	179	164(5.2), 161.1(23.6), 151.1(59.2), 147(19.7), 137.1(44.7), 136.2(21), 133(76.3), 123(39.4), 121.1(31.5), 119.1(76.3), 103(29), 92.9(25), 91(100), 79.2(11.8), 76.9(10.5)	Coniferaldehyde
19.	2.74	371	353(2.4),325(6), 311(3.6), 279(8.4), 251(12), 233(4.8), 219(88), 205(15.6), 191(100), 173(4.8), 163(16.8), 145(3.6), 123(8.4), 85(2.4), 73(3.6)	Kusunokinin

Source: Reproduced from Pandey et al., 2014 with permission from Royal Society of Chemistry.

3.6.1 Identification of Propenyl Phenols and Their Derivatives

Peak 12 (*m/z* 151) and peak 16 (*m/z* 135) were tentatively identified as hydroxychavicol and chavicol, respectively, on the basis of their ESI-MS/MS fragmentation patterns. Base peak ions for chavicol and hydroxychavicol were observed at *m/z* 107 and 123, respectively, due to the loss of ethylene from [M+H]⁺ ions and further elimination of CO molecules generated fragment ions at *m/z* 79 and 95 for chavicol and hydroxychavicol, respectively. Peak 12 (*m/z* 151) showed a favored loss of water resulting in a fragment ion at *m/z* 133 indicating the presence of two vicinal hydroxyl groups, suggesting thereby that it is hydroxychavicol. A further loss of ethylene molecule from the fragment ion at *m/z* 133 generated the fragment ion at *m/z* 105. Hydroxychavicol and chavicol were previously reported from *P. betle*. Peak 10 (*m/z* 165) was identified as eugenol by comparing the retention time and the characteristic fragment ions with the corresponding authentic standard. It was also observed that peak 24 (*m/z* 195), tentatively identified as methoxyeugenol, generated similar types of fragment ions. Fragmentation of [M+H]⁺ ions of 10 and 24 showed loss of methanol resulting in the ions at *m/z* 163 and 133 and ethylene resulting in the ions at *m/z* 167 and 137, respectively. A further loss of ethylene from fragment ions *m/z* 133 and 163 yielded fragment ions at *m/z* 105 and 135 for eugenol and methoxyeugenol, respectively. Fragment ions observed at *m/z* 150 and 180 were due to the loss of methyl radicals from eugenol and methoxyeugenol, respectively. A fragment ion observed at *m/z* 133 was due to the loss of formaldehyde from the fragment ion of methoxyeugenol at *m/z* 163, which proved that peak 24 corresponds to methoxy derivative of eugenol. Loss of formaldehyde is a characteristic fragmentation of methyl aryl ether. These fragmentation patterns were in agreement with the reported literature. Peaks 13 (*m/z* 235) and 17 (*m/z* 207) were identified as allylpyrocatechol-3,4-diacetate and eugenyl acetate, respectively, by comparing their retention time and the characteristic fragment ions with the corresponding authentic standards. Peaks 4 (*m/z* 177) and 23 (*m/z* 193) were tentatively identified as chavicol acetate and allylpyrocatechol monoacetate, respectively, on the basis of the ESI-MS/MS fragmentation pattern, which matched with authentic standards. Fragment ions observed at *m/z* 135, 151, 165 and 193 were due to the loss of neutral ketene molecules from [M+H]⁺ ions of chavicol acetate (*m/z* 177), allylpyrocatechol monoacetate (*m/z* 193), eugenyl acetate (*m/z* 207) and allylpyrocatechol-3,4-diacetate (*m/z* 235), respectively. Loss of acetic acid from the corresponding [M+H]⁺ ions produced fragment ions at *m/z* 117, 133, 147 and 175 for chavicol acetate, allylpyrocatechol monoacetate, eugenyl acetate and allylpyrocatechol-3,4-diacetate, respectively. These compounds were previously reported from *P. betle*. Peaks 20 (*m/z* 209) and 28 (*m/z* 179) were tentatively identified as sinapinaldehyde and coniferal-dehyde, respectively, on the basis of the ESI-MS/MS fragmentation pattern, which is in agreement with the reported literature. Similar fragmentation patterns were observed for both compounds, which indicated that these are analogous. Elimination of methanol, methyl radicals and water from [M+H]⁺ ions of coniferaldehyde and sinapinaldehyde produced fragment ions at *m/z* 147, 177, at *m/z* 164, 194 and at *m/z* 161, 191, respectively. Fragment ions observed at *m/z* 151 and 181 for coniferaldehyde and sinapinaldehyde, respectively, were due to the elimination of CO from [M+H]⁺ ions.

3.6.2 Identification and Characterization of Other Phytoconstituents

Peak 2 (*m/z* 223) was tentatively identified as nerolidol and characterized on the basis of the ESI-MS/MS fragmentation pattern. Fragment ions observed at *m/z* 205 were due to the loss of a water molecule and at *m/z* 181 were due to the loss of a propene moiety from the parent ion. Nerolidol was previously reported from Piper species (*Piper falconeri, Piper guineense, Piper marginatum* and *Piper nigrum*). Three flavonoids, peaks 3 (*m/z* 303), 9 (*m/z* 291) and 11 (*m/z* 301), were tentatively identified as quercetin, catechin, and 8-hydroxy-5,7-dimethoxyflavanone, respectively, and characterized on the basis of the ESI-MS/MS fragmentation pattern, which was previously reported. Fragments formed by retro-Diels–Alder (RDA) reactions are the most diagnostic fragments for flavonoid identification. Fragment ions observed at *m/z* 153, 139, and 197 for quercetin, catechin, and 8-hydroxy-5,7-dimethoxyflavanone, respectively, were due to RDA reaction. Quercetin and catechin were previously reported from *P. betle*, and 8-hydroxy-5,7-dimethoxyflavanone was previously reported from *Piper hispidum*.

Two lignans, peaks 15 (*m/z* 357) and 29 (*m/z* 371), were tentatively identified as pluviatilol and kusunokinin, respectively, and characterized on the basis of the ESI-MS/MS fragmentation pattern. Pluviatilol and kusunokinin were reported from *Piper brachystachyum* and *Piper chaba*, respectively. The base peak ion for pluviatilol observed at *m/z* 205 was due to the loss of $C_7H_8O_2$ followed by the loss of neutral CO molecule and the peak at *m/z* 123 was due to the loss of $C_{13}H_{14}O_4$. The fragment ion observed at *m/z* 339 was due to the loss of a water molecule. The fragment ion observed at *m/z* 219 for kusunokinin was due to the loss of $C_9H_{12}O_2$; a further loss of neutral CO molecule generated a fragment ion at *m/z* 191, which is the base peak ion. The fragment ion observed at *m/z* 353 was due to the loss of a water molecule. The proposed fragmentation of lignans is in agreement with that reported in literature. Peak 21 (*m/z* 355) was tentatively identified as chlorogenic acid and characterized on the basis of the ESI-MS/MS fragmentation pattern. Previously, chlorogenic acid reported from *P. betle* leaf showed anticancer activity. A predominant fragment ion at m/z 163 was observed due to the loss of stable acidic part ($C_7H_{12}O_6$); a further loss of neutral CO molecule generated a fragment ion at m/z 135. The presence of two vicinal hydroxyl groups favored the elimination of water, resulting in a fragment ion at m/z 145. Fragmentation of chlorogenic acid in positive mode is in agreement with the previously reported literature. Peak 19 (m/z 127) was tentatively identified as pyrogallol and characterized on the basis of the ESI-MS/MS fragmentation pattern. Their base peak observed at m/z 99 was due to the loss of CO, which is a characteristic fragmentation of phenols. Further loss of water from the peak at m/z 99 gave a fragment ion at m/z 81. The presence of two vicinal hydroxyl groups on the benzene ring favored the elimination of water resulting in a fragment ion at m/z 109. Peak 22 (m/z 285) was tentatively identified as 3-(2,4,5-Trimethoxyphenyl)-2-acetoxyl-hydroxypropane and characterized on the basis of the ESI-MS/MS fragmentation pattern. 3-(2,4, 5-Trimethoxyphenyl)-2-acetoxy-1-hydroxypropane was previously reported from *P. clusii*. A fragment ion at m/z 225 was observed due to the loss of two formaldehyde molecules; a further loss of acetic acid produced a fragment ion at m/z 165. Elimination of water from the [M+H]$^+$ ion generated a fragment ion at m/z 267, which gave a fragment ion at m/z 207 due to the loss of two formaldehyde molecules.

3.7 VALIDATION PARAMETERS OF QUANTITATIVE ANALYSIS

The UHPLC-MRM method was validated according to the guidelines of International Conference on Harmonization (ICH, Q2R1) with respect to linearity, limit of detection (LOD), limit of quantification (LOQ), accuracy, precision, stability, and recovery.

3.7.1 Linearity, Limit of Detection and Limit of Quantification

A series of concentrations of standard solution were prepared for the establishment of calibration curves. The linearity of calibration was performed by the analytes-to-internal standard (IS) peak area ratios versus the nominal concentration, and the calibration curves were constructed with a weight ($1/x^2$) factor by least squares linear regression. The applied calibration model for all curves was y = ax + b, where y = peak area ratio (analyte/IS), x = concentration of the analyte, a = slope of the curve, and b = intercept. The LOD and LOQ were determined based on calibration curve method by the following equations: LOD = (3.3 × S_{xy})/S_a and LOQ = (10 × S_{xy})/S_a, where S_{xy} is the residual standard deviation of the regression line and S_a is the slope of a calibration curve. The LOD for three analytes varied from 0.13 to 0.48 ng/mL and LOQ from 0.41 to 1.47 ng/mL. All the calibration curves showed good linearity ($r^2 \geq 0.9981$) within the test ranges.

3.7.2 Precision, Stability, and Recovery

The intraday and interday variations, which were chosen to determine the precision of the developed method, were investigated by determining three analytes with IS in six replicates during a single day and by duplicating the experiments on three consecutive days. Variations of the peak area were taken as the measures of precision and expressed as percentage relative standard deviations (% RSD).The overall intraday and interday precision were not more than 1.0%. Stability of sample solutions stored at room temperature was investigated by replicate injections of the sample solution at 0, 2, 4, 8, 12, and 24 h. The RSD values of stability of the three analytes ≤ 2.5%.

A recovery test was applied to evaluate the accuracy of this method. Three different concentration levels (high, middle, and low) of the analytical standards were added into the samples in triplicate, and average recoveries were determined by the following equation: recovery (%) = (observed amount − original amount)/spiked amount × 100%.

The analytical method developed had good accuracy with overall recovery in the range from 96.14% to 98.46% (RSD ≤ 1.6%) for all analytes.

3.8 QUANTITATIVE ANALYSIS

The proposed UHPLC-MRM method was subsequently applied to determine the content of three major bioactive phenolics in the thirteen landraces of *P. betle*. Quantitative analysis of these phenolics showed that allylpyrocatechol-3,4-diacetate was below the detection level in Meetha Patta, Shirpurkata, Assam Paan, Mahoba, and Saufia; similarly, eugenyl acetate was below the detection level in Shirpurkata, Kapoori, Assam Paan, and Mahoba, whereas eugenol was detected in all landraces except Meetha Patta and Kapoori. The content of three phenolics, namely, allylpyrocatechol-3,4-diacetate, eugenyl acetate, and eugenol in thirteen landraces is summarized in Table 3.2.

TABLE 3.2 The content (mg/g) of three analytes in thirteen *Piper betle* landraces

P. BETLE LANDRACES	ALLYLPYROCATECHOL-3,4-DIACETATE	EUGENYL ACETATE	EUGENOL	TOTAL (MG/G)	TOTAL (%)
	ANALYTES MG/G (MEAN ± SD, N = 3)				
Meetha Patta*	Bdl	0.4 ± 0.06	bdl	0.4	0.04
Sanchi*	0.16 ± 0.37	1.5 ± 0.02	1.71 ± 0.04	3.37	0.337
Shirpurkata	Bdl	bdl	0.06 ± 0.11	0.06	0.006
Kapoori	0.03 ± 0.14	bdl	bdl	0.03	0.003
Assam Paan	Bdl	bdl	4.72 ± 0.06	4.72	0.472
Nagpuri	4.25 ± 0.22	56.4 ± 0.25	113.20 ± 0.10	173.85	17.385
Jalesar Green	3.19 ± 0.26	180.0 ± 0.31	230.67 ± 0.14	413.86	41.386
Jagarnathi	0.6 ± 0.02	34.8 ± 0.18	66.80 ± 0.21	102.2	10.22
Deshi	2.2 ± 0.04	133.33 ± 0.24	177.33 ± 0.22	312.86	31.286
Mahoba	bdl	bdl	24.27 ± 0.26	24.27	2.47
Saufia	bdl	34.53 ± 0.61	85.20 ± 0.31	119.73	11.976
Jalesar White	16.8 ± 0.05	90.4 ± 0.59	119.33 ± 0.35	226.53	22.653
Bangladeshi	20.4 ± 0.16	121.3 ± 0.06	190.67 ± 0.05	332.37	33.237

Source: Reproduced from Pandey et al., 2014 with permission from Royal Society of Chemistry.
bdl, Below detection level; *West Bengal.

The results showed remarkable differences in the content of phenolics in all the thirteen landraces of *P. betle*. The total content of three phenolics in Jalesar Green (41.4%) was much higher than other landraces of *P. betle*; other notable landraces with high content were Bangladeshi (33.2%) > Deshi (31.3%) > Jalesar White (22.7%). Eugenol was the most abundant compound in all the landraces and found highest (23.1%) in Jalesar Green. The study showed that the concentration of phenolics varies significantly with the geographical origin of *P. betle*. Therefore, the developed method might be quite suitable and reasonable for quality control of *P. betle*.

3.9 PRINCIPAL COMPONENT ANALYSIS

Principal component analysis (PCA) was applied to establish the correlation and discrimination of *P. betle* landraces using software STATISTICA 7.0. The UHPLC-MS data (including t_R, mass, and % area of peaks) of all the 13 *P. betle* landraces were subjected to PCA using 12 variables (*m/z* 223, 303, 177, 291, 165, 301, 151, 235, 357, 207, 135, 127, 209, 355, 285, 193, 195, 179, and 371). The PC1 vs. PC2 plot clearly brings out the relationship among all the thirteen landraces, which are classified into four distinct groups based on the relative content of identified compounds.

The dimensions of 12 peaks were reduced to three principal components explaining about 72.6% variation. 33.21% variation was explained by PC1, composed of peaks at t_R 1.73 min (*m/z* 151, 235, 357), t_R 1.93 min (*m/z* 127, 209, 355), and t_R 2.34 min (*m/z* 179). The PC2 explained 23.2% variation in *P. betle* landraces. The similarity in Meetha Patta and Nagpuri landraces was due to the compound at t_R 1.5 min (*m/z* 291), as their contribution was highest (20–23%). Landraces Shirpurkata and Saufia were together due to compounds at t_R 1.09 min (*m/z* 223, 303); similarly Deshi, Bangladeshi, and Jalesar White were in the same group due to the similarity in the relative abundance of compounds at t_R (1.09, 1.5, 1.81, and 2.34 min). The compounds at t_R 1.09 min (*m/z* 223, 303) were present in the highest abundance (23–47%), whereas compounds at t_R 1.5 min (*m/z* 291), t_R 1.81 min (*m/z* 135, 207), and t_R 2.34 min (*m/z* 179) were present in equal amounts (6–7%). Sanchi, Kapoori, Assam Paan, Jalesar Green, Mahoba, and Jagnathi were very similar due to the high abundance of compounds at t_R 1.09 min (*m/z* 223, 303), which was 60% in Assam Paan and more than 70% in Jalesar Green, Mahoba, and Jagnathi. The compound at t_R 1.5 min (*m/z* 291) was present in 10–12% abundance in Assam Paan, Jalesar Green, and Jagnathi but completely absent in Sanchi, Kapoori, and Mahoba landraces.

3.10 ANTIMICROBIAL ACTIVITY

The in vitro antimicrobial activity of ethanolic leaf extracts of *P. betle* landraces, Nagpuri, Jalesar Green, Jagarnathi, Deshi, Mahoba, Saufia, Jalesar White and Bangladeshi, which showed high content of major bioactive phenolics in quantitative analysis were tested against four bacteria namely, *Escherichia coli* (Ec, ATCC 9637), *Pseudomonas aeruginosa* (Psu, ATCC BAA-427), *Staphylococcus aureus* (Sa, ATCC 25923) and *Klebsiella pneumoniae* (Kpn, ATCC 27736) and six fungi namely, *Candida albicans* (Ca, patient isolate), *Cryptococcus neoformans* (Cn, patient isolate, *Sporothrix schenckii* (Ss, patient isolate), *Trichophyton mentagrophytes* (Tm, patient isolate), *Aspergillus fumigatus* (Af, patient isolate) and *Candida parapsilosis* (Cp, ATCC-22019). The susceptibility testing was performed by Standard Broth Microdilution method as per Clinical and Laboratory Standard Institute (CLSI) guidelines using RPMI 1640 medium buffered with MOPS [3-(N-morpholino) propanesulfonic acid] for fungal cultures and Mueller Hinton Broth (Difco) for bacterial cultures in 96 well microtiter plates. The maximum concentration tested was 500 µg/mL and the inoculums load in each test well was in the range of $1-5 \times 10^3$ cells. The plates were incubated for 24–48 h for yeasts, 72–96 h for mycelial fungi at 35°C and 24 h for bacteria at 37°C and read visually as well as spectrophotometrically (SpectraMax) at 492 nm for determination of minimal inhibitory concentrations (MICs).

3.11 ANTIMICROBIAL ACTIVITY EVALUATION

All the ethanolic leaf extracts tested exhibited antifungal activity where Jalesar Green, which showed higher content of quantified phenolics was best against *C. parapsilosis* and *C. neoformans* with MIC values 6.25 and 31.2 µg/mL, respectively (Table 3.3). Other notable landraces, Bangladeshi, Jalesar White and Deshi, were effective against *C. albicans*, *C. parapsilosis* and *C. neoformans* with MIC values ranging from 15.6 to 31.2 µg/mL (Table 3.3). However, none of the extracts were found to be active against bacteria at even 500 µg/mL concentration.

TABLE 3.3 Minimum inhibitory concentration (MIC) in µg/mL of ethanolic leaf extracts of *Piper betle* landraces against bacteria and fungi

P. betle LANDRACES	MIC IN µG/ML AGAINST									
	BACTERIA				FUNGI					
	1	2	3	4	5	6	7	8	9	10
Nagpuri	>500	>500	>500	>500	250	125	500	500	>500	125
Jalesar Green	>500	>500	>500	>500	62.5	31.2	250	250	500	6.25
Jagarnathi	>500	>500	>500	>500	62.5	62.5	250	250	500	125
Deshi	>500	>500	>500	>500	31.2	15.6	125	125	125	31.2
Mahoba	>500	>500	>500	>500	62.5	62.5	500	500	>500	125
Saufia	>500	>500	>500	>500	125	125	250	500	>500	125
Jalesar White	>500	>500	>500	>500	31.2	15.6	125	125	250	31.2
Bangladeshi	>500	>500	>500	>500	62.5	31.2	125	125	500	62.5
Standards										
Gentamycin	12.5	3.12	6.25	6.25	nd	nd	nd	nd	nd	nd
Norfloxacin	0.024	0.78	0.39	0.05	nd	nd	nd	nd	nd	nd
Fluconazole	nd	nd	nd	nd	1	2	2	>32	>32	2
Amphotericin B	nd	nd	nd	nd	0.02	0.13	0.3	0.25	0.5	0.02

Source: Reproduced from Pandey et al., 2014 with permission from Royal Society of Chemistry.
1, *Escherichia coli* (ATCC 9637); 2, *Pseudomonas aeruginosa* (ATCC BAA-427); 3, *Staphylococcus aureus* (ATCC 25923); 4, *Klebsiella pneumoniae* (ATCC 27736); 5, *Candida albicans;* 6, *Cryptococcus neoformans;* 7, *Sporothrix schenckii;* 8, *Trichophyton mentagrophytes;* 9, *Aspergillus fumigates;* 10, *Candida parapsilosis* (ATCC-22019); nd, not detectable.

Conclusions

4

A new approach to characterize population diversity in the form of landraces, gender or geographical location–based differences in the plants was developed by combination of direct analysis in real time (DART) time of flight mass spectrometric (MS) method and chemometric analysis. This method is simple, rapid and provides a useful qualitative fingerprint, which could be used for quality control of PB. The DART-MS of the PB leaves could be recorded without any sample preparation. The abundances of the characteristic phenols were different in the landraces. Based on exact mass measurements, the elemental compositions of 17 constituents were determined in the phytochemical screening. Fifteen compounds (pyrogallol, chavicol, estragole, allylpyrocatechol, chavibetol, phenylalanine, chavicol acetate, coniferaldehyde, allylpyrocatechol acetate, β-caryophyllene, chavibetol acetate, sinapinaldehyde, nerolidol, allylpyrocatechol diacetate and 8-hydroxy-5,7-dimethoxyflavanone) were identified in intact PB leaf. Principal component analysis (PCA) could easily classify the PB landraces into four groups depending on their positions in PCA plot. The peaks at m/z 150, 151, 164, 165, 175 193, 207, 235 and 252 are common to all male and female landraces and hence constitutive for PB. Based on peak abundance in PB leaf, the identification of sex (gender) of unknown PB (Sirugamil, Calcutta Bangla, Helisa, Shirpurkata, Khasi and Malvi) could be successfully done by comparison with known male (Kapoori Chintalpudi) and female (Gachi). Factor analysis (FA)/PCA also showed the expected grouping of the landraces. Based on relative abundance of three bioactive compounds, namely, allylpyrocatechol, chavibetol and allylpyrocatechol acetate, PB can be separated into two broad groups of low therapeutic and high therapeutic potential. The DART-MS profile for allylpyrocatechol and chavibetol is a good predictor of their relative content, which can be correlated to their therapeutic potential. Selection of suitable landrace for a particular drug candidate is also possible by comparing the mass fingerprints based on the bioactive content. Thus, this method benefits from its simplicity and the potential extensibility for other metabolomic based plant studies.

A simple, fast, sensitive, and reliable ultrahigh performance liquid chromatography–electrospray ionization–tandem mass spectrometric method was also developed and validated for identification and structural characterization of phytoconstituents. Simultaneous determination of major bioactive phenolics namely, allylpyrocatechol-3, 4-diacetate, eugenyl acetate, and eugenol in thirteen landraces of *P. betle* leaf extracts were completed successfully. A total of nineteen compounds were identified and characterized and schematic fragmentation pathways for all the identified compounds were proposed. This method was also applied to investigate the major bioactive phenolics content in thirteen PB landraces showed remarkable differences in their content. Thus, this method provides an excellent tool for quality assessments of PB due to its high capacity, sensitivity, selectivity and shorter analysis time. In vitro antimicrobial activity of only those landraces, which showed the high content of bioactive phenolics was also evaluated. This approach will go a long way in exploring plant wealth for judicious utilization of plant wealth for human wellness.

References

Abdul Ghani, Z. D. F., A. H. Ab Rashid, K. Shaari, and Z. Chik. "Urine NMR metabolomic study on biochemical activities to investigate the effect of *P. betle* extract on obese rats." *Applied Biochemistry and Biotechnology* 189 (2019): 690–708.

Abrahim, N. N., M. S. Kanthimathi, and A. Abdul-Aziz. "*Piper betle* shows antioxidant activities, inhibits MCF-7 cell proliferation and increases activities of catalase and superoxide dismutase." *BMC Complementary and Alternative Medicine* 12, no. 1 (2012): 220.

Adhikary, P., J. Banerji, D. Chowdhury, A. K. Das, C. Deb, S. R. Mukherjee, and A. Chatterjee. "Antifertility effect of *Piper betle* Linn. extract on ovary and testis of albino rats." *Indian Journal of Experimental Biology* 27, no. 10 (1989): 868–870.

AICEP. Summary report of the All India Coordinated Ethno-biological Project (AICEP) of Ministry of Environment & Forests, Govt, of India. (1994).

Al-Adhroey, A. H., Z. M. Nor, H. M. Al-Mekhlafi, A. A. Amran, and R. Mahmud. "Antimalarial activity of methanolic leaf extract of *Piper betle* L." *Molecules* 16, no. 1 (2011): 107–118.

Amonkar, A. J., M. Nagabhushan, A. V. D'souza, and S. V. Bhide. "Hydroxychavicol: a new phenolic antimutagen from betel leaf." *Food and Chemical Toxicology* 24, no. 12 (1986): 1321–1324.

Anthropological Survey of India. People of India Project Report for 1994. (1994).

Arambewela, L., M. Arawwawala, K. G. Kumaratunga, D. S. Dissanayake, W. D. Ratnasooriya, and S. P. Kumarasingha. "Investigations on *Piper betle* grown in Sri Lanka." *Pharmacognosy Reviews* 5, no. 10 (2011): 159.

Arambewela, L., M. Arawwawala, and W. D. Ratnasooriya. "Antidiabetic activities of aqueous and ethanolic extracts of *Piper betle*. leaves in rats." *Journal of Ethnopharmacology* 102, no. 2 (2005a): 239–245.

Arambewela, L., M. Arawwawala, and W. D. Ratnasooriya. "Antinociceptive activities of aqueous and ethanol extracts of *Piper betle*. leaves in rats." *Pharmaceutical Biology* 43, no. 9 (2005b): 766–772.

Arawwawala, M., L. Arambewela, and W. D. Ratnasooriya. "Gastroprotective effect of *Piper betle* Linn. leaves grown in Sri Lanka." *Journal of Ayurveda and Integrative Medicine* 5, no. 1 (2014): 38.

Arias, T., R. C. Posada, and A. Bornstein. "New combinations in Manekia, an earlier name for Sarcorhachis (Piperaceae)." *Novon: A Journal for Botanical Nomenclature* 16, no. 2 (2006): 205–209.

Bajad, S., K. L. Bedi, A. K. Singla, and R. Johri. "Antidiarrhoeal activity of piperine in mice." *Planta Medica* 67, no. 03 (2001a): 284–287.

Bajad, S., K. L. Bedi, A. K. Singla, and R. Johri. "Piperine inhibits gastric emptying and gastrointestinal transit in rats and mice." *Planta Medica* 67, no. 02 (2001b): 176–179.

Bajpai, V., D. Sharma, B. Kumar, and K. Madhusudanan. "Profiling of *Piper betle* Linn. cultivars by direct analysis in real time mass spectrometric technique." *Biomedical Chromatography* 24, no. 12 (2010): 1283–1286.

Bajpai, V., R. Pandey, M. P. S. Negi, K. H. Bindu, N. Kumar, and B. Kumar. "Characteristic differences in metabolite profile in male and female plants of dioecious *Piper betle* L." *Journal of Biosciences* 37, no. 1 (2012a): 1061–1066.

Bajpai, V., R. Pandey, M. P. S. Negi, N. Kumar, and B. Kumar. "DART MS based chemical profiling for therapeutic potential of *P. betle* landraces." *Natural Product Communications* 7, no. 12 (2012b): 1627–1629.

Balasubrahmanyam, V. R., J. K. Johri, A. K. S. Rawat, R. D. Tripathi, and R. S. Chaurasia. "Betelvine (*P. betle* L.)." Economic Botany Information Service, National Botanical Research Institute, Lucknow, India (1994).

Baviskar, H. P., G. T. Dhake, M. A. Kasai, N. B. Chaudhari, and T. A. Deshmukh. "Review of *Piper betle*." *Research Journal of Pharmacognosy and Phytochemistry* 9, no. 2 (2017): 128.

Bhattacharya, S., D. Banerjee, A. K. Bauri, S. Chattopadhyay, and S. K. Bandyopadhyay. "Healing property of the Piper betel phenol, allylpyrocatechol against indomethacin-induced stomach ulceration and mechanism of action." *World Journal of Gastroenterology: WJG* 13, no. 27 (2007): 3705.

Bhattacharya, S., M. Subramanian, S. Roychowdhury, A. K. Bauri, J. P. Kamat, S. Chattopadhyay, and S. K. Bandyopadhyay. "Radioprotective property of the ethanolic extract of *Piper betel* leaf." *Journal of Radiation Research* 46, no. 2 (2005): 165–171.

Bissa, S., D. Songara, and A. Bohra. "Traditions in oral hygiene: chewing of betel (*Piper betle* L.) leaves." *Current Science* 92, no. 1 (2007): 26–28.

Chang, M. C., B-J. Uang, C. Y. Tsai, H-L. Wu, B-R. Lin, C. S. Lee, Y-J. Chen et al. "Hydroxychavicol, a novel betel leaf component, inhibits platelet aggregation by suppression of cyclooxygenase, thromboxane production and calcium mobilization." *British Journal of Pharmacology* 152, no. 1 (2007): 73–82.

Chen, S. J., B-N. Wu, J-L. Yeh, Y-C. Lo, I. S. Chen, and I-J. Chen. "C-fiber-evoked autonomic cardiovascular effects after injection of *Piper betle* inflorescence extracts." *Journal of Ethnopharmacology* 45, no. 3 (1995): 183–188.

Chopra, V. L, R. A. Vishwakarma. *Plants for Wellness and Vigour*. New India Publishing Agency, New Delhi, India, 2018, p. 399. ISBN: 9789386546036.

Choudhary, D., and R. K. Kale. "Antioxidant and non-toxic properties of Piper betle leaf extract: in vitro and in vivo studies." *Phytotherapy Research: An International Journal Devoted to Pharmacological and Toxicological Evaluation of Natural Product Derivatives* 16, no. 5 (2002): 461–466.

Chu, N-S. "Effects of betel chewing on the central and autonomic nervous systems." *Journal of Biomedical Science* 8, no. 3 (2001): 229–236.

CSIR (Council of Scientific and Industrial Research, New Delhi). The Wealth of India, 8: 84–94. CSIR, New Delhi (1969).

Dasgupta, N., and B. De. "Antioxidant activity of *Piper betle* L. leaf extract in vitro." *Food Chemistry* 88, no. 2 (2004): 219–224.

Daware, M. B., A. M. Mujumdar, and S. Ghaskadbi. "Reproductive toxicity of piperine in Swiss albino mice." *Planta Medica* 66, no. 03 (2000): 231–236.

Desai, S. K., V. Gawali, A. B. Naik, and L. L. D'souza. "Potentiating effect of piperine on hepatoprotective activity of *Boerhaavia diffusa* to combat oxidative stress." *International Journal of Pharmacology* 4 (2008): 393–397.

Dhuley, J. N., P. H. Raman, A. Mujumdar, and S. R. Naik. "Inhibition of lipid peroxidation by piperine during experimental inflammation in rats." *Indian Journal of Experimental Biology* 31, no. 5 (1993): 443–445.

Dorman, H. J. D., and S. G. Deans. "Antimicrobial agents from plants: antibacterial activity of plant volatile oils." *Journal of Applied Microbiology* 88, no. 2 (2000): 308–316.

Evans, P. H., W. S. Bowers, and E. J. Funk. "Identification of fungicidal and nematocidal components in the leaves of *Piper betle* (Piperaceae)." *Journal of Agricultural and Food Chemistry* 32, no. 6 (1984): 1254–1256.

Fazal, F., P. P. Mane, M. P. Rai, K. R. Thilakchand, H. P. Bhat, P. S. Kamble, P. L. Palatty, and M. S. Baliga. "The phytochemistry, traditional uses and pharmacology of *Piper betel*. Linn (Betel Leaf): A pan-asiatic medicinal plant." *Chinese Journal of Integrative Medicine* (2014): 1–11.

Ferreres, F., A. P. Oliveira, A. Gil-Izquierdo, P. Valentão, and P. B. Andrade. "*Piper betle* leaves: profiling phenolic compounds by HPLC/DAD-ESI/MSn and anticholinesterase activity." *Phytochemical Analysis* 25, no. 5 (2014): 453–460.

Frodin, D.G. "History and Concepts of Big Plant Genera" *Taxon* 53, no. 3 (2004): 753–776.

Ganguly, S., S. Mula, S. Chattopadhyay, and M. Chatterjee. "An ethanol extract of *Piper betle* Linn. mediates its anti-inflammatory activity via down-regulation of nitric oxide." *Journal of Pharmacy and Pharmacology* 59, no. 5 (2007): 711–718.

Ghani, Z. D. F. A., J. M. Husin, A. H. Ab Rashid, K. Shaari, and Z. Chik. "Biochemical studies of *Piper betle* L leaf extract on obese treated animal using 1H-NMR-based metabolomic approach of blood serum samples." *Journal of Ethnopharmacology* 194 (2016): 690–697.

Ghosh, K., and T. Bhattacharya. "Chemical constituents of *Piper betle* Linn. (Piperaceae) roots." *Molecules* 10, no. 7 (2005): 798–802.

Gilani, A. H., N. Aziz, I. M. Khurram, Z. A. Rao, and N. K. Ali. "The presence of cholinomimetic and calcium channel antagonist constituents in *Piper betle* Linn." *Phytotherapy Research: An International Journal Devoted to Pharmacological and Toxicological Evaluation of Natural Product Derivatives* 14, no. 6 (2000): 436–442.

Gopalan, C., B.V. Ramasastri, and S.C. Balasubramanian. Nutritive Value of Indian Foods, p. 108. National Institute of Nutrition (ICMR), Hyderabad, India (1984).

Gragasin, M. C. B., A. M. Wy, B. P. Roderos, M. A. Acda, and A. D. Solsoloy. "Insecticidal activities of essential oil from *Piper betle* Linn against storage insect pests." *Philippine Agriculturist Scientist* 89, no. 3 (2006): 212–216.

Guha, P. "Betel leaf: the neglected green gold of India." *Journal of Human Ecology* 19, no. 2 (2006): 87–93.

Guha, P. and R.K. Jain. Status Report On Production, Processing and Marketing of Betel Leaf (*Piper betle* L.). Agricultural and Food Engineering Department, IIT, Kharagpur, India (1997).

Gundala, S. R., and R. Aneja. *"Piper betel* leaf: a reservoir of potential xenohormetic nutraceuticals with cancer-fighting properties." *Cancer Prevention Research* 7, no. 5 (2014): 477–486.

Guo, N., K. Ablajan, B. Fan, H. Yan, Y. Yu, and D. Dou. "Simultaneous determination of seven ginsenosides in Du Shen Tang decoction by rapid resolution liquid chromatography (RRLC) coupled with tandem mass spectrometry." *Food Chemistry* 141, no. 4 (2013): 4046–4050.

Gupta, S., S. M. Gupta, A. P. Sane, and N. Kumar. "Chlorophyllase in *Piper betle* L. has a role in chlorophyll homeostasis and senescence dependent chlorophyll breakdown." *Molecular Biology Reports* 39, no. 6 (2012): 7133–7142.

Hajare, R., V. M. Darvhekar, A. Shewale, and V. Patil. "Evaluation of antihistaminic activity of piper betel leaf in guinea pig." *African Journal of Pharmacy and Pharmacology* 5, no. 2 (2011): 113–117.

Héberger, K.. "Chemoinformatics—multivariate mathematical–statistical methods for data evaluation." In *Medical Applications of Mass Spectrometry*, pp. 141–169. Elsevier, Amsterdam, 2008.

Jane, N. S., A. P. Deshmukh, and M. S. Joshi. "Review of study of different diseases on betelvine plant and control measure." *International Journal of Application or Innovation in Engineering and Management* 3, no. 3 (2014): 560–563.

Kanjwani, D., T. P. Marathe, S. Chiplunkar, and S. Sathaye. "Evaluation of immunomodulatory activity of methanolic extract of *Piper betel.*" *Scandinavian Journal of Immunology* 67, no. 6 (2008): 589–593.

Kaveti, B., L. Tan, K. T. S. Sarnnia, and M. Baig. "Antibacterial activity of *Piper betle* leaves." *International Journal of Pharmacy Teaching and Practices* 2, no. 3 (2011): 129–132.

Khajuria, A., N. Thusu, U. Zutshi, and K. L. Bedi. "Piperine modulation of carcinogen induced oxidative stress in intestinal mucosa." *Molecular and Cellular Biochemistry* 189, no. 1–2 (1998): 113–118.

Kumar, N. "Betelvine (*Piper betle* L.) cultivation: a unique case of plant establishment under anthropogenically regulated microclimatic conditions." *Indian Journal of History of Science* 34, (1999): 19–32.

Kumar, A., B. R. Garg, G. Rajput, D. Chandel, A. Muwalia, I. Bala, and S. Singh. "Antibacterial activity and quantitative determination of protein from leaf of Datura stramonium and *Piper betle* plants." *Pharmacophore* 1, no. 3 (2010a): 184–195.

Kumar, N., P. Misra, A. Dube, S. Bhattacharya, M. Dikshit, and S. Ranade. "*Piper betle* Linn. a maligned Pan-Asiatic plant with an array of pharmacological activities and prospects for drug discovery." *Current Science* 99, no. 7 (2010b): 922–932.

Kumar N., S. Gupta, A. N. Tripathi. Gender-specific responses of Piper betle L. to low temperature stress: Changes in chlorophyllase activity. *Biol. Plant.* 50, (2006): 705–708.

Lee, S. A., S. S. Hong, X. H. Han, J. S. Hwang, G. J. Oh, K. S. Lee, M. K. Lee, B. Y. Hwang, and J. S. Ro. "Piperine from the fruits of Piper longum with inhibitory effect on monoamine oxidase and antidepressant-like activity." *Chemical and Pharmaceutical Bulletin* 53, no. 7 (2005): 832–835.

Lei, D., C-P. Chan, Y-J. Wang, T-M. Wang, B-R. Lin, C-H. Huang, J-J. Lee, H-M. Chen, J-H. Jeng, and M-C. Chang. "Antioxidative and antiplatelet effects of aqueous inflorescence *Piper betle* extract." *Journal of Agricultural and Food Chemistry* 51, no. 7 (2003): 2083–2088.

Majumdar, B., S. G. R. Chaudhuri, A. Ray, and S. K. Bandyopadhyay. "Effect of ethanol extract of *Piper betle* Linn leaf on healing of NSAID—induced experimental ulcer-A novel role of free radical scavenging action." *Indian Journal of Experimental Biology* 41, no. 4 (2003): 311–315.

Medicinal Plants in China: A Selection of 150 Commonly Used Species. WHO Regional Publications: Western Pacific Series No. 2, 1989.

Misra, K. H., N. Ranjita, K. Ramu, and M. Bandyopadhyay. "Evaluation of antiasthmatic effect of ethanol extract of *Piper betle* Linn. against histamine induced bronchospasm in guinea pigs." *International Journal of Basic and Applied Chemical Sciences* 4, no. 1 (2014): 67–73.

Mohottalage, S., R. Tabacchi, and P. M. Guerin. "Components from Sri Lankan *Piper betle* L. leaf oil and their analogues showing toxicity against the housefly, Musca domestica." *Flavour and Fragrance Journal* 22, no. 2 (2007): 130–138.

Mujumdar, A. M., J. N. Dhuley, V. K. Deshmukh, P. H. Raman, and S. R. Naik. "Antiinflammatory activity of piperine." *Japanese Journal of Medical Science and Biology* 43, no. 3 (1990): 95–100.

Nagori, K, M. K. Singh, A. Alexander, T. Kumar, D. Dewangan, H. Badwaik, and D. K. Tripathi. "*Piper betle* L.: a review on its ethnobotany, phytochemistry, pharmacological profile and profiling by new hyphenated technique DART-MS (Direct Analysis in Real Time Mass Spectrometry)." *Journal of Pharmacy Research* 4, no. 9 (2011): 2991–2997.

Nath, T. K., M. Inoue, F. E. Pradhan, and M. H. Kabir. "Indigenous practices and socio-economics of *Areca catechu* L. and *Piper betel* L. based innovative agroforestry in northern rural Bangladesh." *Forests, Trees and Livelihoods* 20, no. 2–3 (2011): 175–190.

Nayar, M. P., and A. R. K. Sastry. "*Red data book of Indian plants.*" Botanical Survey of India, Kolkata, 1987.

Panda, S., and A. Kar. "Dual role of betel leaf extract on thyroid function in male mice." *Pharmacological Research* 38, no. 6 (1998): 493–496.

Panda, S., M. Sikdar, S. Biswas, R. Sharma, and A. Kar. "Allylpyrocatechol, isolated from betel leaf ameliorates thyrotoxicosis in rats by altering thyroid peroxidase and thyrotropin receptors." *Scientific Reports* 9, no. 1 (2019): 1–12.

Panda, S., R. Sharma, and A. Kar. "Antithyroidic and hepatoprotective properties of high-resolution liquid chromatography–mass spectroscopy-standardized *Piper betle* leaf extract in rats and analysis of its main bioactive constituents." *Pharmacognosy Magazine* 14, no. 59 (2018): 658.

Pandey, R., P. Chandra, M. Srivastva, K. R. Arya, P. K. Shukla, and B. Kumar "A rapid analytical method for characterization and simultaneous quantitative determination of phytoconstituents in *P. betle* landraces using UPLC-ESI-MS/MS." *Analytical Methods* 6, no. 18 (2014): 7349–7360.

Paranjpe, R., S. R. Gundala, N. Lakshminarayana, A. Sagwal, G. Asif, A. Pandey, and R. Aneja. "*Piper betel* leaf extract: anticancer benefits and bio-guided fractionation to identify active principles for prostate cancer management." *Carcinogenesis* 34, no. 7 (2013): 1558–1566.

Parmar, V. S., S. C. Jain, K. S. Bisht, R. Jain, P. Taneja, A. Jha, O. D. Tyagi et al. "Phytochemistry of the genus Piper." *Phytochemistry* 46, no. 4 (1997): 597–673.

Parmar, V. S., S. C. Jain, S. Gupta, S. Talwar, V. K. Rajwanshi, R. Kumar, A. Azim et al. "Polyphenols and alkaloids from Piper species." *Phytochemistry* 49, no. 4 (1998): 1069–1078.

Patra, B., M. T. Das, and S. K. Dey A review on *Piper betle* L. *Journal of Medicinal Plants* 4, no. 6 (2016): 185–192.

Piper betle. The Wealth of India: The dictionary of Indian raw material and industrial products. Council of Scientific and Industrial Research (CSIR), New Delhi (2003).

Prabhu, M. S., K. Platel, G. Saraswathi, and K. Srinivasan "Effect of orally administered betel leaf (*Piper betle* Linn.) on digestive enzymes of pancreas and intestinal mucosa and on bile production in rats." *Indian Journal of Experimental Biology* 33, no. 10 (1995): 752–756.

Prabu, S. M., M. Muthumani, and K. Shagirtha. "Protective effect of *Piper betle* leaf extract against cadmium-induced oxidative stress and hepatic dysfunction in rats." *Saudi Journal of Biological Sciences* 19, no. 2 (2012): 229–239.

Pradhan, D., K. A. Suri, D. K. Pradhan, and P. Biswasroy. "Golden heart of the nature: *Piper betle* L." *Journal of Pharmacognosy and Phytochemistry* 1, no. 6 (2013):147–167.

Priya, S. G.. "Studies on morphological characterization, variability, heritability and genetic advance in betelvine (*Piper betle* Linn.)." Phd Diss., Andhra Pradesh Horticultural University, 2011.

Quijano-Abril, M. A., R. Callejas-Posada, and D. R. Miranda-Esquivel. "Areas of endemism and distribution patterns for Neotropical *Piper species* (Piperaceae)." *Journal of Biogeography* 33, no. 7 (2006): 1266–1278.

Rai, D. R., V. K. Chourasiya, S. N. Jha, and O. D. Wanjari. "Effect of modified atmospheres on pigment and antioxidant retention of betel leaf (*Piper betel* L.)." *Journal of Food Biochemistry* 34, no. 5 (2010): 905–915.

Ramji, N., N. Ramji, R. Iyer, and S. Chandrasekaran. "Phenolic antibacterials from *Piper betle* in the prevention of halitosis." *Journal of Ethnopharmacology* 83, no. 1–2 (2002): 149–152.

Ranade, S. A., A. Soni, and N. Kumar. "SPAR profiles for the assessment of genetic diversity between male and female landraces of the dioecious betelvine plant (Piper betle L.)." In *Ecosystems Biodiversity*, p. 443. IntechOpen, 2011.

Ranade, S., A. Verma, M. Gupta, and N. Kumar. "RAPD profile analysis of betel vine cultivars." *Biologia Plantarum* 45, no. 4 (2002): 523–527.

Rathee, J. S., B. S. Patro, S. Mula, S. Gamre, and S. Chattopadhyay. "Antioxidant activity of *Piper betel* leaf extract and its constituents." *Journal of Agricultural and Food Chemistry* 54, no. 24 (2006): 9046–9054.

Rawat, A. K. S., U. Shome, and V. R. Balasubrahmanyam. "Analysis of the volatile constituents of Piper betle L. cultivars—a chemosystematic approach." In *Medicinal and Poisonous Plants of the Tropics: Proceedings of Symposium 5–35 of the 14th International Botanical Congress*, Berlin, 24 July–1 August 1987/AJM Leeuwenberg (compiler). Wageningen [Netherlands]: Pudoc, 1987.

Rawat, A. K. S., R. D. Tripathi, A. J. Khan, and V. R. Balasubrahmanyam. "Essential oil components as markers for identification of *Piper betle* L. cultivars." *Biochemical Systematics and Ecology* 17, no. 1 (1989): 35–38.

Rekha, V. P. B., M. Kollipara, B. R. S. S. Gupta, Y. Bharath, and K. K. Pulicherla. "A review on *Piper betle* L.: Nature's promising medicinal reservoir." *American Journal of Ethnomedicine* 1, no. 5 (2014): 276–289.

Rimando, A. M., B. H. Han, J. H. Park, and M. C. Cantoria. "Studies on the constituents of Philippine *Piper betle* leaves." *Archives of Pharmacal Research* 9, no. 2 (1986): 93–97.

Row, L-C. M., and J-C. Ho. "The antimicrobial activity, mosquito larvicidal activity, antioxidant property and tyrosinase inhibition of *Piper betle*." *Journal of the Chinese Chemical Society* 56, no. 3 (2009): 653–658.

Samantaray, S., A. Phurailatpam, A. K. Bishoyi, K. A. Geetha, and S. Maiti. "Identification of sex-specific DNA markers in betel vine (*Piper betle* L.)." *Genetic Resources and Crop Evolution* 59, no. 5 (2012): 645–653.

Sarkar, M., P. Gangopadhyay, B. Basak, K. Chakrabarty, J. Banerji, P. Adhikary, and A. Chatterjee. "The reversible antifertility effect of *Piper betle* Linn. on Swiss albino male mice." *Contraception* 62, no. 5 (2000): 271–274.

Sarkar, D., P. Saha, S. Gamre, S. Bhattacharjee, C. Hariharan, S. Ganguly, R. Sen, et al. "Anti-inflammatory effect of allylpyrocatechol in LPS-induced macrophages is mediated by suppression of iNOS and COX-2 via the NF-κB pathway." *International Immunopharmacology* 8, no. 9 (2008): 1264–1271.

Sengupta, R., and J. K. Banik. "A review on betel leaf (pan)." *International Journal of Pharmaceutical Sciences and Research* 4, no. 12 (2013): 4519.

Shah, S. K., G. Garg, D. Jhade, and N. Patel. "*Piper betle*: phytochemical, pharmacological and nutritional value in health management." *International Journal of Pharmaceutical Sciences Review and Research* 38 (2016): 181–189.

Sharma, K. K., R. Saikia, J. Kotoky, J. Kalita, and J. Das. "Evaluation of antidermatophytic activity of *Piper betle*, Allamanda cathertica and their combination: an in vitro and in vivo study." *International Journal of Pharm Tech Research* 3 (2011): 644–651.

Seetha Lakshmi, B. and K. C. Naidu. "Comparative Morphoanatomy of *Piper betle* L. Cultivars in India," *Annals of Biological Research* 1, no. 2 (2010): 128–134.

Shitut, S., V. Pandit, and B. K. Mehta. "The antimicrobial efficiency of *Piper betle* Linn leaf (stalk) against human pathogenic bacteria and phytopathogenic fungi." *Central European Journal of Public Health* 7, no. 3 (1999): 137–139.

Singh, M., S. Shakya, V. K. Soni, A. Dangi, N. Kumar, and S-M. Bhattacharya. "The n-hexane and chloroform fractions of *Piper betle* L. trigger different arms of immune responses in BALB/c mice and exhibit antifilarial activity against human lymphatic filarid *Brugia malayi*." *International Immunopharmacology* 9, no. 6 (2009): 716–728.

Sripradha, S. "Betel leaf-the green gold." *Journal of Pharmaceutical Sciences and Research* 6, no. 1 (2014): 36–37.

Sunila, E. S., and G. Kuttan. "Immunomodulatory and antitumor activity of Piper longum Linn and piperine." *Journal of Ethnopharmacology* 90, no. 2–3 (2004): 339–346.

Suryasnata, D., I. S. Sandeep, R. Parida, S. Nayak, and S. Mohanty. "Variation in volatile constituents and Eugenol content of five important betelvine (*Piper betle* L.) landraces exported from eastern India." *Journal of Essential Oil Bearing Plants* 19, no. 7 (2016): 1788–1793.

Toprani, R., and D. Patel. "Betel leaf: Revisiting the benefits of an ancient Indian herb." *South Asian Journal of Cancer* 2, no. 3 (2013): 140–141.

Trakranrungsie, N., A. Chatchawanchonteera, and W. Khunkitti. "Ethnoveterinary study for antidermatophytic activity of *Piper betle, Alpinia galanga* and *Allium ascalonicum* extracts in vitro." *Research in Veterinary Science* 84, no. 1 (2008): 80–84.

Tripathi, R., G. Khare, and N. Kumar. "Effect of low temperature stress on betle vine (*P. betle* L.) types, Bangla and Desavari". *Journal of Spices and Aromatic Crops* 9, no. 2 (2000): 141–143.

Tripathi, S., N. Singh, S. Shakya, A. Dangi, S. Misra-Bhattacharya, A. Dube, and N. Kumar. "Landrace/gender-based differences in phenol and thiocyanate contents and biological activity in *Piper betle* L." *Current Science* 91, no. 6 (2006): 746–749.

Valentão, P., R. F. Gonçalves, C. Belo, P. G. de Pinho, P. B. Andrade, and F. Ferreres. "Improving the knowledge on *Piper betle*: targeted metabolite analysis and effect on acetylcholinesterase." *Journal of Separation Science* 33, no. 20 (2010): 3168–3176.

Varunkumar, V. S., G. N. Mali, S. Joseph, and N. O. Varghese. "Evaluation of the anticariogenic effect of crude extract of *Piper betle* by assessing its action on salivary pH–an in vitro study." *Journal of Dental and Medical Sciences* 13, no. 8 (2014): 43–48.

Venkadeswaran, K., A. R. Muralidharan, T. Annadurai, V. V. Ruban, M. Sundararajan, R. Anandhi, P. A. Thomas, and P. Geraldine. "Antihypercholesterolemic and antioxidative potential of an extract of the plant, *Piper betle*, and its active constituent, eugenol, in triton WR-1339-induced hypercholesterolemia in experimental rats." *Evidence-Based Complementary and Alternative Medicine* (2014): 11. https://doi.org/10.1155/2014/478973.

Verma, A., N. Kumar, and S. Ranade. "Genetic diversity amongst landraces of a dioecious vegetatively propagated plant, betelvine (*Piper betle* L.)." *Journal of Biosciences* 29, no. 3 (2004): 319–328.

Vyawahare, N. S., and S. L. Bodhankar. "Neuropharmacological profile of *Piper betel* leaves extract in mice." *Pharmacologyonline* 2 (2007): 146–162.

Wirotesangthong, M., N. Inagaki, H. Tanaka, W. Thanakijcharoenpath, and H. Nagai. "Inhibitory effects of *Piper betle* on production of allergic mediators by bone marrow-derived mast cells and lung epithelial cells." *International Immunopharmacology* 8, no. 3 (2008): 453–457.

Index

A

Aam, 2
Abaxial, 7
Abundance, 27, 33, 34, 37, 40, 45, 46, 47, 53, 57
AccuTOF, 23
Acetonitrile, 41, 42, 43, 44
3β-Acetyl ursolic acid, 9
Acquisition rate, 26
Acrid, 10
Aegle marmelos, 2
Algae, 1
Aliquot, 42
Allylbenzene, 9
Allyl diacetoxy benzene, 9
Allylguaiacol, 9
4-Allyl-phenol, 9
Allylpyrocatechol, 4, 9, 12, 13, 15, 27, 28, 57
Allylpyrocatechol acetate, 28, 57
Allylpyrocatechol diacetate, 9, 28, 57
Allylpyrocatechol-3, 4-diacetate, 42, 44, 46, 48, 51, 52, 58
Allylpyrocatechol monoacetate, 9, 47, 48
Aluva, 21
Aluva southern, 21
Amino acids, 6
Amphotericin B, 55
Ampro odekkali local local paan, 21
Ampro odekkali local-1, 21
Ampro odekkali local medicinal paan, 21
Analgesic, 12
Anerobes, 4, 12
Anethole, 9
Anginosus, 11
Anthelmintic, 10
Anthropogenic, 5, 7, 39
Antiallergic, 12, 13
Antiamoebic, 12
Antiasthmatic, 17
Antibacterial, 11, 12

Anticancer, 12, 14, 50
Anticariogenic, 11
Antidermatophytic, 16
Antidiabetic, 11, 12
Antifertility, 12, 13
Antifungal, 11, 12, 54
Antihypercholesterolemic, 16
Antihyperglycemic, 12
Antiinfective, 12
Antiinflammatory, 12, 13, 17
Antileishmanial, 12
Antimalarial, 11, 13
Antimicrobial, 10, 12, 30, 41, 54, 58
Antimutagenic, 12
Antimutation, 11
Antinociceptive, 11, 16, 17
Antioxidant, 12, 14, 32
Antiplatelet, 11, 12, 15
APC, 27, 29, 33, 36, 37, 38, 39
Aphrodisiac, 4, 10, 11
Appetite, 11
Areca catechu, 2
Arecoline, 9
Aroma, 4, 7
Aromatic, 5, 10
Ascorbic acid, 9
Aspergillus fumigates, 54, 55
Assam pan, 20, 22, 41, 51, 52, 53
Asthma, 11, 17
Ayurved, 11
Azadirachta indica, 2

B

Bacteria, 10, 11, 12, 54, 55
Bactericidal, 12
Banana, 2
Banyan, 2
Bareja, 7, 8
Barley, 2
Baroi kamakhya, 22

Barouj, 7
BEH, 43, 44
Bengali, 3
Betelvine, 3
Betel quid, 4
Beverage, 2
Bhang, 2
Bheet, 7, 8
Bilva, 2
Biosynthesis, 39
Bitter, 10
Bodeh, 3
Botanical, 2, 5
Brain, 11
Breast, 12
Breeding, 5
Bronchitis, 11
Bundle, 7

C

Cabe, 3
CAD, 43
Cadinene, 9
Calcium, 6, 15
Calibration, 23, 26, 43, 50, 51
Calotropis gigantea, 2
Camphene, 9
Candida albicans, 54, 55
Candida parapsilosis, 54, 55
Cannabis indica, 2
Carang, 3
Carbohydrates, 6
Cardamom, 10
Cardiovascular, 12
Carminative, 4, 10, 11
Carotene, 6, 9
Carvacrol, 9, 27, 36
β-Caryophyllene, 27, 28, 36, 57
Caryophylline, 9
Cascabela thevetia, 2
Catarrh, 12
Catechin, 45, 49
Cepharadione, 9
Chandan, 2
Chavibetol, 2, 9, 16, 27, 28, 29, 36, 39, 57
Chavibetol acetate, 9, 27, 28, 36, 39, 57
Chavicol, 9, 28, 46, 48, 57
Chavicol acetate, 28, 45, 48, 57
Chemopreventive, 14
Chest, 10
Chlorogenic acid, 47, 50

Chlorophyll, 4, 9
Chlorophyllase, 9
Cholinomimetic, 14, 15
Chronic, 10
CHV, 27, 29, 36, 37, 39
1,8-Cineole, 9
Citrobacter freundii, 12
Citrobacter koseri, 12
Climber, 4, 5
Clove, 10
Cluster analysis, 24
Coconut, 2, 8
Cocus nucifera, 2
Coffee, 3
Commensal, 10
Commercial, 2, 32
Concentration, 17, 36, 39, 42, 50, 51, 54, 55
Coniferaldehyde, 47, 49, 57
Constipation, 11
Contamination, 10
Contraceptive, 11, 13
Cotton, 2
Cough, 11
Cryptococcus neoformans, 54, 55
Crystals, 7
CSH, 44
Cultivation, 4, 7, 39
Cultural, 1, 2, 3
Curcuma longa, 2
Curve, 16, 43, 50, 51
Cynodon dactylon, 2
Cytotoxic, 11

D

Datura, 2
Datura stramonium, 2
Dental, 10
Dermatophytosis, 12
Desiccative, 10
Desmostachia bipinnata, 2
Detector voltage, 26
Detoxication, 11
Dicots, 2
Digestion, 7
Digestive, 4, 6, 7
2,4-Dihydroxy-allylbenzene, 9
Dioecy, 9, 31, 32
4-Dimethoxy-allylbenzene, 9
Diphtheria, 11
Discharge electrode, 26

DNA, 9, 16, 33
Doorva, 2
Dotriacontanoic acid, 9
Dyspnea, 11

E

Eardrops, 11
E. coli, 55
Ecological, 38, 39
Economic, 2, 6
Elaeocarpus ganitrus, 2
Elements, 7, 26
Elephantiasis, 11
Elimination, 32, 48, 49, 50
Endogenous, 10
Enterococcus faecalis, 12
Enzymes, 6
Epidermal cells, 7
Epidermis, 7
Epiphytes, 39
Escherichia coli, 12, 54, 55
Estragole, 9, 27, 28, 36, 57
Ethanol, 20, 42
Euclidean, 24
Eugenol, 9, 16, 27, 28, 33, 36, 38, 42, 44, 45,
    48, 51, 52, 53, 58
Eugenol acetate, 9
Eugenol methyl ether, 9
Eugenyl acetate, 38, 42, 44, 46, 48, 51,
    52, 58
Exhilarant, 10
Expectorant, 10
Extraction, 20, 42
Eye, 11

F

FA, 24, 37, 57
Factor analysis, 24, 57
Fat, 5, 12
Febrifuge, 10
Fever, 11
Fiber, 1, 6, 16
Ficus bengalensis, 2
Ficus religiosa, 2
Fingerprints, 20, 57
Flora, 10
Fluconazole, 55
Folklore, 6
Formic acid, 41, 43, 44
Fragmentation, 23, 25, 48, 49, 50, 58

Freetree, 24
Fungus, 11
Furrows, 7

G

Gas beam temperature, 26
Gastritis, 11
Gastroprotective, 11, 17
Gender, 9, 24, 25, 31, 32, 33, 34, 35, 39, 57
Gentamycin, 55
Giant sequoia, 1
Glucose, 12
Gonorrhea, 11
Gossipium, 2
Green gold, 4
Grid electrode, 26
Gudhal, 2
Guduchi, 2
Gujarati, 3

H

Haldi, 2
Halitosis, 4, 10, 11, 12
Harsingar, 2
Healing, 11, 12, 13, 15
Heart, 11
Helium flow-rate, 26
Hepatoprotective, 12
Herbivores, 39
Hibiscus rosa sinensis, 2
Highland, 7
Hindi, 3
Hordeum vulgare, 2
HRMS, 27
Hudang, 3
Humidity, 7, 8, 39
Hydroxycatechol, 9
Hydroxychavicol, 9, 13, 15, 27, 46, 48
8-Hydroxy-5,7-dimethoxytlavanone, 45,
    49, 57
3-Hydroxy-4-methoxyallylbenzene, 9
4-Hydroxy-3-methoxy-allylbenzene, 9
Hygiene, 10
Hy leaves, 22
Hysteria, 11

I

Idioblasts, 7
Immunomodulatory, 12, 14, 16

Indigestion, 11
Indomethacin, 12
Infant, 11
Inflammation, 11
Insecticidal, 11, 14
Instrumentation, 43
Interventions, 7
Ion mode, 26
Ion trap, 43
Iron, 6
Isoeugenol, 9

J

Jaccard coefficient, 24
Jakun, 3
Javanese, 3
Jerak, 3
JMS-T100LC, 23

K

Kaker, 9
Kamal, 2
Kaner, 2
Kenayek, 3
Kerakap, 3
Klebsiella pneumoniae, 12, 54, 55
K-means, 24, 33
Kusha, 2
Kusunokinin, 47, 49

L

Lactobacillus acidophilus, 11
Laxative, 10
Leucorrhea, 11
Limonene, 9
Linearity, 50, 51
Liver, 11
Lod, 50, 51
Longevity, 7
Loq, 50, 51
Lung, 11

M

Madar, 2
Magnesium, 6
Malay, 3

Malayalam, 7
Mangifera indica, 2
Manures, 7
Marathi, 3
Marigold, 2
Mass center, 23
Mass range, 26, 39, 43
Masticatory, 7
Matrix, 24, 36
Medicinal, 1, 37
Melayu, 3
Metabolite, 10, 19, 26, 27, 31, 32, 36
Methanol, 12, 14, 20, 23, 44, 49
4-Methoxy-allylbenzene, 9
2-Methoxy-4-allyl-phenol, 9
Methoxyeugenol, 47, 48
Methyl eugenol, 9, 36
Methyl piperbetol, 9
Microbial, 10
MICs, 54
Midrib, 6, 7
Mimetic, 12
Minerals, 5, 6
Minitab, 24
Mitis, 11
3-(N-Morpholino) propanesulfonic
    acid, 54
M-nitrobenzyl alcohol, 23
Monocots, 2
Monsoon, 7
Mouth, 10
MRM, 43, 50, 51
Mukhbhushan, 3
Multicellular tector, 7
Multivariate, 24, 40
Musa, 2
Mutagenic, 9
Mutan, 10
Mutation, 5

N

Nagarbael, 3
Nagavallari, 3
Nagavallika, 3
Nagbael, 3
Nagballi, 3
Nagini, 3
Needle voltage, 26
Neem, 2

Neighbor joining, 24
Nelumbo nucifera, 2
Nerolidol, 27, 45, 49, 57
Neuropharmacological, 15
Neutrophils, 13
Niacin, 6
NJ, 24, 36
N-nitrosomorpholine, 9
N-nitrosopiperidine, 9
N-nitrosopyrrolidine, 9
Node, 7
Norfloxacin, 55
Nyctanthes arbortristis, 2

O

Obligate, 4, 5, 12
Ocimum sanctum, 2
Odor, 11
Oil, 7, 9, 11, 13, 14
Ointment, 12
Optimization, 25, 26, 44
Oral, 4, 10, 12
Organoleptic, 4, 10
Orifice 1 potential, 26
Orifice 1 temperature, 26
Orifice 2 potential, 26
Origin, 10, 38, 40, 53

P

Paleoherbs, 2
P-allyl-anisole, 9
P-allyl-phenol, 9
Parenchymatous, 7
Parijat, 2
PCA, 24, 25, 29, 30, 31, 32, 34, 36, 37, 40,
    53, 57
P-Cymene, 9
PDA, 42, 43
Peak voltage, 26
Peepal, 2
Pelu, 3
Peperomia, 3
Pepper, 3, 7
Percolator, 20
pH, 11
Pharmaceutical, 2
Phenols, 9, 27, 32, 35, 37, 44, 48, 50, 57
Phenyl alanine, 27, 28

Phylogenetic, 2
Pinene, 9
Piper brachystachyum, 49
Piper chaba, 2, 49
Piper chavya, 21
Piper falconeri, 49
Piper guineense, 49
Piper longum, 2
Piper marginatum, 49
Piper nigrum, 2, 49
Piperaceae, 1, 2, 3
Piperbetol, 2, 9
Piperine, 2, 9
Piperitol, 9
Piperlonguminine, 9
Piperol A, 9
Piperol B, 9
Pippali, 3, 7
Plaque, 11
Platelet, 15
Pluviatilol, 46
Polyethylene glycol, 23
Polymorphic, 24, 33, 36
Potential, 6, 13, 15, 16, 20, 23, 24, 25, 26, 37,
    38, 40, 43, 57
Prana, 10
Precipitation, 39
Precision, 51
Precursor, 43, 44, 45, 46, 47
Principal component analysis, 24, 32, 34,
    53, 57
Product, 32, 43, 44
Propagation, 4, 5
Propenyl phenols, 44, 48
Prosopis cineraria, 2
Protein, 5
Proteus vulgaris, 12
Pseudomonas aeruginosa, 12, 54, 55
Pyrogallol, 46, 50, 57

Q

Quercetin, 45, 49

R

Radioprotective, 16
RAPD, 33
Recovery, 50, 51
Redwoods, 1

Residues, 42
Resolution, 26, 27, 43, 44
Respiratory, 11, 15
RF ion guide potential, 26
Rheumatism, 11
Riboflavin, 6
Ring lens potential, 26
Rudraksha, 2

S

Safrole, 9
Sakai, 3
Salivarius, 11
Salivary, 11
Sampling time, 26
Sandal, 2
Sanskrit, 3, 5
Santalum album, 2
Saptashira, 3, 5
Season, 7
Semang, 3
Sensitivity, 9, 26, 44, 58
Serasa, 3
Sesamum indicum, 2
Sexual, 5, 11
Shami, 2
Sheath, 7
Simulative, 4
Sinapinaldehyde, 27, 46, 49, 57
Sirih, 3
Sirihcina, 3
β-Sitosterol, 9
β-Sitosteryl palmitate, 9
S. mutans, 11
Sporothrix schenckii, 54, 55
Stability, 4, 50, 51
Standard, 10, 27, 42, 43, 44, 48, 50, 51,
    54, 55
Staphylococcus aureus, 12, 54, 55
Statistica, 24, 53
Stearaldehyde, 9
Stearic acid, 9
Stock, 23, 42, 43
Stomachic, 10, 11
Stomatal, 7
Streptococcus mutans, 10
Streptococcus pyrogenes, 12
Styptic, 11
Sugars, 9

Supari, 2
Supernatant, 42
Suruh, 3
Survival, 1, 39
Sweet, 10
Swellings, 11
Synergistic, 10

T

Tagetes, 2
Tamalapaku, 3
Tambool, 3
Tamil, 3
Tannins, 9
Tea, 3
Teeth, 11
Telugu, 3
Temperature, 7, 8, 20, 25, 26, 39, 42, 43,
    44, 51
Terpenes, 2, 9, 27
Terpenoids, 9
α-Terpine, 9
α-Terpinol, 9
Terpinyl acetate, 9
Terrestrial, 1
Thai, 3
Therapeutic, 1, 13, 15, 20, 24, 35, 37, 38, 57
Thiamine, 6
Throat, 10, 11
Thyroid, 17
Til, 2
Tinospora cordifolia, 2
Tracheoids, 7
Tract, 7
Trichome, 7
Trichophyton mentagrophytes, 54, 55
3-(2,4,5-Trimethoxyphenyl)-
    2-acetoxy-l-
    hydroxypropane, 47, 50
Tritriacontane, 9
Tulsi, 2
Tumours, 12

U

UHPLC-ESI-MS/MS, 42, 43, 45, 46, 47
Ulcers, 11
Unani, 11
Ursonic acid, 9

V

Vaginal, 11
Validation, 24, 25, 30, 33, 34, 50
Varnalata, 3
Vascular plants, 1
Vegetative, 2, 4, 5, 32, 33
Vessel, 7
Vetrilai, 3
Vettila, 7
Vitamin, 6
Volatile, 10

Voucher, 20
Vulnerary, 11

W

Ward's method, 24
WHO, 1
Wounds, 11

X

Xeromorphic, 4, 7

Printed in the United States
by Baker & Taylor Publisher Services